AF328050

Expert System Applications in Materials Processing and Manufacturing

Proceedings of a symposium sponsored by the TMS Shaping & Forming Committee, the TMS Synthesis & Analysis in Materials Processing Committee and the ASM-MSD Computer Simulation Activity, held at the TMS Annual Meeting in Anaheim, California, February 18-22, 1990.

Edited by

M.Y. Demeri
Ford Motor Company
Dearborn, Michigan

A Publication of

A Publication of The Minerals, Metals & Materials Society
420 Commonwealth Drive
Warrendale, Pennsylvania 15086
(412) 776-9024

Printed in the United States of America
Library of Congress Catalog Number 89-63259
ISBN Number 0-87339-120-9

© 1990

Preface

This book is a pre-conference publication which contains the completed papers presented at the TMS Annual Meeting Symposium entitled "Expert System Applications in Materials Processing and Manufacturing" held in Anaheim, California, February 18-22, 1990.

The symposium is sponsored by three committees: the TMS Shaping & Forming Committee, the TMS Synthesis & Analysis in Materials Processing Committee and the ASM-MSD Computer Simulation Activity . The organizers of this symposium and their affiliations are as follows:

> Dr. M.Y. Demeri, TMS Shaping & Forming Committee
> Dr. H. Henein and Dr. W. Pardee, TMS SAMP Committee
> Dr. R. Harrison, ASM-MSD Computer Simulation Activity

The symposium has 26 participants. Presentations are divided into four sessions presided by the following session chairmen:

> Session I: Dr. M. Demeri, Research Staff, Ford Motor Co.
> Session II: Dr. N. Karam, Spire Corporation
>   Mr. P. MacNeille, Ford Motor Co.
> Session III: Dr. H. Henein, University of Alberta
>   Dr. W. Pardee, Rockwell International
> Session IV: Dr. R. Harrison, U.S. Army Materials Lab.

The book covers a broad range of the most up-to-date expert system applications in areas such as real time process monitoring, process control, diagnostics, process scheduling and material selection. In the area of shaping and forming, the book contains papers on the use of expert systems in welding, casting and mechanical processing of materials. An editorial review standard, which emphasizes content and quality, is used in selecting and arranging the articles published in this book.

The editor wishes to recognize the co-organizers for their individual contribution towards the "rounding up" of completed manuscripts. R. Harrison contributed three manuscripts, W. Pardee contributed one manuscript and H. Henein contributed one manuscript towards the publication of this book.

The editor wishes to thank the co-organizers and the session chairmen for all the effort they put in preparing for the symposium. Thanks are also due to all participants, especially those authors who took the extra time and effort to prepare full manuscripts for this publication.

M. Y. Demeri
Ford Motor Company
Manufacturing Systems Department
Research Staff
Dearborn, Michigan

October 20, 1989

Table of Contents

PROSPECTS FOR INCORPORATION OF ARTIFICIAL INTELLIGENCE

TECHNIQUES INTO MATERIALS SCIENCE RESEARCH

Ralph J. Harrison

U.S. Army Materials Technology Laboratory
Watertown, MA 02172

ABSTRACT

We will review requirements, progress and prospects for the implementation
of artificial intelligence techniques which can benefit the materials
scientist in carrying out research. We are not proposing to replace the
researcher by an expert system; the BACON program for making scientific
deductions would, if it had existed at the time, not have threatened Francis
Bacon's place in the history of science. We will describe AI tools which
have become available that can be useful additions to the toolkit of
materials research. These may include for example the utilization of
pattern recognition or neural net techniques in addition to more conventional statistical analysis to discover correlations among various phe-
nonema. Knowledge based software for the materials researcher (KBMR) can go
beyond algorithmic evaluation of data and help the researcher in problem
solving. Some of this software must be specific to the requirements of
materials research, while some may be applicable to generic scientific or
engineering research.

Expert System Applications in
Materials Processing and Manufacturing
Edited by M.Y. Demeri
The Minerals, Metals & Materials Society, 1989

<u>Introduction</u>

This paper describes how artificial intelligence techniques could be used to aid the materials researcher in carrying out materials research. In order to avoid overselling AI one possibly should start with a disclaimer that the paper is not going to describe how the materials researcher can be replaced by an intelligent computer program, but rather that many of the things that occupy the researcher's time at present might in the future be done by such a program. It is recognized that research is not simply an application of known techniques and rules to a problem, but it also involves innovation and discovery. AI programs have made some progress in dealing with discovery; there exists an ingenious program referred to as BACON (1-2) which is designed to examine collections of individual scientific facts and to deduce 'scientific laws' from them, perhaps emulating the way Francis Bacon himself applied the 'scientific method'. I should not imagine that the authors of the program would suggest that if their program had been available at the time Bacon lived, Bacon would have hung up his shingle and decided that there was nothing for him to contribute to science. However it is possible that some of the AI discovery tools could have made him even more productive. The focus of this paper on AI is as a tool among other research tools, that can provide a powerful and unique addition to the researcher's 'toolkit'.

<u>Recent Work</u>

Recent work has touched on some of the ways that AI methods can contribute to scientific research. In a previous TMS symposium on AI in Materials Science, I reviewed (3) some of the ways in which AI techniques might be utilized in carrying out simulations in materials research. Abelson et al. (4) have discussed the use of intelligence in scientific computing, showing how intelligent computational tools for scientists and engineers can autonomously prepare simulation experiments from high-level specifications of physical models. These tools can "actively monitor numerical and physical experiments...interpret experimental data and formulate numerical results in qualitative terms...enable their human users to control computational experiments in terms of high-level behavioral descriptions." They envision an 'engineer's assistant' that can begin with a description of a mechanism and then generate efficient numerical programs to predict the dynamical behavior. So far they have explored the development of this tool for engineering dynamical systems, with application to the design of intelligent automatic control systems; for chemical systems with application to characterization of complex reaction mechanisms; and for interpreting results in planetary dynamics. It has been recognized that the mass of data produced by supercomputers requires the development of sophisticated <u>visualization</u> techniques to help a researcher in interpretation. Conventionally these visualization techniques have been directly used by the human researcher, but the AI tool which Abelson et al. have developed automatically outputs the results of the visualization program to computer programs which then perform an analysis on the spatial patterns and print out an interpretation in direct analogue to the way humans might analyze these patterns (4). This work may be regarded as an important step toward the implementation of a futuristic prediction (which was made for chemistry but could just as well refer to a goal for materials research) by John Sculley (5), "AI will boost simulations and hypermedia to new levels of realism and usefulness. For example, we will move from building molecules into two and three dimensional space, to building the environment in which they combine- where each molecule understands the structure and behavior of the others."

    In other work which seems directly applicable as a research tool for
materials problems, Hardt and Chen (6) have shown how AI methods of qualitative simulation might be combined with dimensional analysis techniques to
contribute to understanding of scientific problems, taking simple mechanics
experiments for illustration.

    R. Albrecht, in an examination of possible future applications of AI in
astronomy, wrote (7) of the utility of having a computer-based Research
Assistant, of senior undergraduate or junior graduate level, who can be
talked to in domain terminology (jargon), can be asked to look up material
in the library, who should "provide unsolicited input if relevant,and to
check our reasoning". He stated that AI will enable "building and the
refinement of models, the checking of concepts for consistency, or the
inferencing from interdisciplinary assertions".

    The symposium proceedings (8) in which Albrecht's article appears,
"Knowledge-Based Systems in Astronomy" has many other examples which are of
interest outside the field of astronomy. One of these is a discussion of a
proposal preparation expert system, which certainly can save much of a
researcher's time even without directly aiding in research. Other expert
systems were devised to select astronomical instrument configuration and
exposure time. There is a system to access a hypertext glossary of
technical terms relating to a particular space telescope; another system for
monitoring telemetry to see if problems have developed; an astronomical data
analysis system which has already processed over 2 million astronomical
images, where the expert system has to make decisions on such matters as
whether the image is that of one of a class of interesting objects, where
the selection criteria, the particular algorithm to use in analysis, and how
to define significance level. Classification systems have not only to put
new images in some pre-existing category, but also to "provide measures
indicating any differences between the properties of the object and the
typical properties of the objects in the box" (9). Expert systems for the
classification of galaxies are also described which involve pattern
recognition and diagnosis using classification reasoning. The materials
researcher can certainly find problems which are direct analogues of the
cited problems from astronomy.

    Buchanan et al.(10) describe an expert system to operate upon test data
generated by a simulation of a particle accelerator, which deduced rules for
diagnosing and correcting misalignment problems in the particle beam.
Subsequently a second expert system is then constructed, utilizing the rules
deduced by the first expert system, which could then perform the function of
diagnosing misalignment in functioning particle beams. Again, a materials
experimentalist might very well use an analogous system of simulation
assisted inductive learning for a materials research problem.

Neural Network Applications

    I include a few remarks about possible future applications of neural
networks to materials research although their domain is separate from,
but overlapping with, AI. One obvious application would take advantage of
their known ability in pattern recognition and classification. At this
meeting M. A. Przystupa (11) describes such an application to the prediction
of yield strength from composition. One can predict more extensive use of
neural nets to correlate properties with structure and composition, perhaps
utilizing one neural network for the classification of features by direct
computer processing of raw metallurgical data, including microscope images,
and then utilizing a second neural network to correlate these features with
properties. A more general application of neural network concepts might be

in replacing or supplementing conventional expert systems, using the neural
net weighting factors to represent probabilistic concepts in application of
rules to be implemented in a naturally parallel structure.

KBMR, <u>a</u> Knowledge Based System for Materials Research

I should like now to describe some work done in collaboration with
I.Hulthage, who was one of the developers of the ALADIN expert system
(12-15) for aluminum alloy design. We have tried to examine how a system
could be devised that would be a specific development tool for materials
research. We call the generic system, KBMR, a Knowledge Based System for
Materials Research, a research assistant for materials researchers (16-18).
Some preliminary code has been written to illustrate how some of our ideas
might be implemented in practice, but we consider that the primary
accomplishments so far have been to list important specifications for such a
system. These deal with such things as how it might interact with the
Materials Researcher, what types of autonomous actions it might take, how
the system would represent materials knowledge and how it would acquire,
expand and modify its data base of such knowledge (both through interaction
with the Materials Researcher and with new experimental information and
literature) and how it would evaluate the relevance of its knowledge to the
research goals of the Researcher.

Plan Representation - General

One of the most important specifications for the KBMR system is that it
must be able, not only to represent the <u>goals</u> of the researcher, but also to
represent the actual <u>plan</u> of research. The plan is a central part of any
research program, expecially one which involves a coordinated effort of a
group of researchers. With respect to the even more generic problem of plan
representation we may cite a very germane recent discussion by Fiksel and
Hayes-Roth (19) devoted to the subject of planning and of how interactive
knowledge systems can "support humans engaged in the learning and thinking
stages of planning ...to assist by analyzing, monitoring, and refining plans
initiated by humans." The emphasis is in knowledge browsing and hypothesis
testing, rather than in search procedures. In their analysis they list four
requirements of a system for aiding planning: knowledge base maintenance,
plan maintenance and validation, qualitative logical analysis, and plan
monitoring and revision. They have implemented their methodology in the
LIBRA system, which performs functions of knowledge representation and
maintainance, with provisions for reasoning and inferencing, and for tracing
accountability for each item of knowledge. It permits continuous
modifications to the conceptual model of the domain. The authors (19)
summarize their concept of a knowledge-based system for interactive planning
by saying it should "provide a flexible and accessible repository for the
shared knowledge of various participants...a systematic organizing structure
for the large amount of qualitative and quantitative information developed
...facilitate rapid response to unforeseen changes in plan assumptions,
outcome of actions, or external conditions."

Plan Representation in KBMR

The cited summary (19) also includes many of the requirements of the
KBMR system as we have seen it. The method for implementing plan
representation which we consider appropriate for the types of plans
important to research programs, and which we have considered for the KBMR,
is to represent plans by a sequence of simulation models, either qualitative

4

(20), semiquantitative (21) or quantitative. A goal which is a required property of the material, or more generally the required behavior under some specified dynamical conditions, is directly represented in a simulation of these conditions. Subgoals may involve requirements for attainment of certain values of parameters in the simulation. In turn the evaluation of the parameters may involve their determination through a separate simulation, or they may be determined through other information sources in the knowledge base. The plan may specify tentative sequences of steps designed to move toward the goal and possibly other directions to take if certain anticipated or unanticipated difficulties are found to block progress. By incorporating a research plan representable as a dynamical model, the KBMR can keep reporting back to the Materials Researcher whether new information acquired represents progress toward the goal or is an indication that snags are developing. It can continually update its evaluation of the optimum path through a network where the weights keep changing as a result of new information acquired, and it will also be making decisions on what new information to try to acquire. The Materials Researcher can intervene and modify the plan and corresponding model if the results obtained seem to be counterintuitive, but can restore the original model if it is then decided that intuition should not be trusted. The KBMR system may be valuable, not only in helping an individual researcher, but in coordinating the efforts of a group of researchers working on a large common project. The plan representation will simultaneously represent the various subtasks of each researcher, and at the same time provide means to cue the individual researcher with information on how information gathered by others is affecting all parts of the plan.

Materials Data Representation

With respect to the representation of materials knowledge by the KBMR, the problem is more than how to represent what is usually termed materials data, although the system does need to have access to materials data bases. Much important work is being carried out in the materials community (22-24) aimed at increasing the ease of computer access to various materials data bases, including methods of querying these bases from expert systems (24). Many properties depend upon knowledge of microstructural details, in turn which are influenced by preparation procedures which can be so varied as to defy conventional tabulation. To facilitate reasoning about properties on the basis of microstructure, one requires an adequate representation of these features. There has been a start in this in the incorporation into the ALADIN system of the Hornbogen (25) classification of such metallic structures as grains, grain boundaries, phase topology, voids, cracks, etc. There is need for extension of this representation in many directions even apart from the obvious extension to increase the data base to cover micro-structural features of other types of materials than represented by aluminum alloys in the ALADIN system.

Microstructural Representation – Scale or Granularity Problem

There is also the more fundamental question raised by the problem of 'scale'; that is, for example if the microstructural features are looked at in the atomic scale, there are infinitely more varieties of geometrical arrangements which are classified only statistically on the larger scale, and the atomistic properties may influence the microstructural features in a way unique to the individual atomic interactions, both static and dynamic. The time scale is also one which plays a role, in that there may be rapid thermal motions which are only visible when one examines very small time intervals, and there may also be some slow diffusive motions which influence the macroscopic properties over times many orders of magnitude larger. For mechanical properties such as plastic deformation, fracture, etc., a detailed description of the mechanism may involve actions going on over multiple time and distance scales.

The problem involved here is a familiar one, for example spoken of as the granularity problem by Hobbs (26), which he suggested one deal with by "constructing simple theories out of more complex ones", that is, complex phenomena are to be organized as a set of computationally tractable local theories. The links between these local theories are formed by what he called _articulation_ processes. Hobbs illustrates his ideas by noting that a highway may be described as a one-dimensional object on a roadmap, but as two-dimensional for the purpose of traffic control, and as three-dimensional by highway construction engineers.

An example of how the problem of granularity might impinge upon a materials scientist or a KBMR is the structure of, say, a polycrystalline ceramic. The grains might be recognized by their appearance in a metallurgical section where in the optical microscope, a grain boundary appears as a one dimensional trace of a two dimensional surface defined by the discontinuity in orientation between two crystallites. At one level of description one may construct a model for, say, small elastic deformations of the ceramic, in which the crystallites are regarded as simply small regions of an anisotropic elastic continuum with orientation axes discontinuous at the boundaries. Border conditions for the elasticity problem would ensure that there is no relative slip at the boundaries. A coarser level of description might ignore the polycrystalline nature of the ceramic and try to represent it by a continuum with some average effective elastic properties. The previous model might be used to calculate the appropriate effective constants. However at times one might have to go to a model at a level containing more detail in which the atomic structure will modify the continuum elastic treatment of the crystallites.

Representation of Dynamical Processes in Materials

The knowledge representation problem for the materials research often requires dynamical processes to be modeled, as well as static structures. The general problem of modeling time-varying states is a difficult one in AI. It is possibily somewhat simpler for the materials problem where one can often represent the process through a simulation involving a known general form for the equation of change, and the dependence on initial and external conditions. Again, as has been pointed out in connection with the representation of the research plan, one can use qualitative or semiquantitative simulations when the quantitative equations are not known or are too difficult to solve exactly. The problems of scale also crop up in dynamical process representation, since evolution of macrostructure is implicitly dependent on the evolution of the microstructure, but often the dynamical equations for each scale may be set up independently, with perhaps some implicit consistency conditions to be observed. The definition of the state of the system in itself affects the dynamical description, since as on one scale the system might be in a single time invariant state, while on a more microscopic scale the system is seen as an ensemble of dynamically changing microstates.

<u>KBMR</u> - <u>Progress</u> <u>and</u> <u>Future</u> <u>Development</u>

In addition to the work already described, we have constructed a simple experimental prototype of KBMR (18) in order to partly emulate a fictitious dialog between the researcher and the System described in reference (17). The purpose of this dialog is for KBMR to debrief the researcher on the progress made recently in the particular research in order to incorporate this new information in the KBMR knowledge base, and also to let the researcher know about relevant related work contained in its prior knowledge base or retrieved from self-directed literature searches. It is hoped that future work will provide a more complete prototype proof of the KBMR concept which can then be used to justify the very extensive commitment of resources which would be necessary to make it into a system that would really demonstrate the impressive potential ability of AI tools in aiding materials research.

References

1. P. Langley, "BACON.1: A general discovery system,"<u>Proceedings of the Second National Conference of the Canadian Society for Computational Studies in Intelligence</u>,pp.173-180,(1978)
2. P. Langley, G.L. Bradshaw and H.A. Simon, "Rediscovering Chemistry with the BACON System",in <u>Machine Learning: An Artificial Intelligence Approach</u>, R.S.Michalski,J.G. Carbonell, and T.M. Mitchell, Eds.,Tioga, Palo Alto, Calif.(1983)
3. R.J. Harrison, "AI and Simulation in Materials Science Research", in <u>Artificial Intelligence Applications in Materials Science</u>, R.J. Harrison and L.D. Roth, Eds., Metallurgical Society of AIME, Warrendale,PA,pp.1-13(1986)
4. H. Abelson, M. Eisenberg,M. Halfant, J. Katzenelson, E. Sacks, G.J. Sussman, J. Wisdom, K. Yip, "Intelligence in Scientific Computing", Comm. of the ACM, $\underline{32}$, 546-562 (1989)
5. John Sculley, "The Relationship Between Business and Higher Education: a Perspective on the 21st Century", Comm. of the ACM, $\underline{32}$, 1056-1061(1989)
6. S.L. Hardt and G. Chen, "Semi-intuitive computations and artificial intelligence:Dimensional analysis as a case study",Computers in Physics $\underline{3}$, 55-62 (1989)
7. R. Albrecht, "Applications of AI in Astronomy: A view towards the Future", in A. Heck and F. Murtagh,(Eds.) <u>Knowledge-Based</u> <u>Systems</u> <u>in</u> <u>Astronomy</u> Springer (1989)p.247-258
8. A. Heck and F. Murtagh,(Eds.) <u>Knowledge-Based</u> <u>Systems</u> <u>in</u> <u>Astronomy</u> Springer (1989)
9. Michael J. Kurtz, "Classification and Knowledge", p.91-106, in A. Heck and F. Murtagh,(Eds.) <u>Knowledge-Based</u> <u>Systems</u> <u>in</u> <u>Astronomy</u> Springer (1989)
10. B. Buchanan, J. Sullivan, T. Cheng, S. Clearwater (1988), "Simulation Assisted Inductive Learning", in, <u>Proc.</u> <u>7th</u> <u>National</u> <u>Conference</u> <u>on</u> <u>Artificial Intelligence</u>
11. M.A. Przystupa, "Predictions of Yield Strength from Composition Using Neural Networks", <u>this</u> <u>proceedings</u>.
12. I. Hulthage, M.L. Farinacci, M.S. Fox, M.D. Rychener, "Knowledge Based Alloy Design", Ch. 6, in <u>Expert Systems for Engineering Design</u> , Academic Press, Boston,MA,(1988)
13. I. Hulthage, M. Przystupa, M.L. Fox, M.D. Rychener,"The Representation of Metallurgical Knowledge for Alloy Design", <u>Artificial</u> <u>Intelligence</u> <u>for</u> <u>Engineering</u> <u>Design</u>, <u>Analysis</u> <u>and</u> <u>Manufacturing</u> (AI EDAM) 1(3), (1988)
14. M.L. Farinacci, M.S. Fox, I. Hulthage, M.D. Rychener, "The development of ALADIN, an expert system for Aluminum alloy design",in <u>Third International Conference on Advanced Information Technology</u>. North-Holland, Amsterdam, November 1986 (Also Carnegie-Mellon technical report CMU-RI-TR-86-5)

15. M.D. Rychener, M.L. Farinacci, I. Hulthage, M.S. Fox, "Integrating mult-iple knowledge sources in ALADIN, an alloy design system",in Fifth national conference on artificial intelligence, AAAI,(1986),p.878

16. R.J. Harrison, I. Hulthage, "Use of Artificial Intelligence Techniques in Materials Research: Application to Research on High-Strength Steels", in Proceedings 34th Sagamore Army Materials Conference 1987, Plenum Press

17. R.J. Harrison, I. Hulthage, "Applications of Knowledge Based Techniques to Materials Research", in Proceedings of Army Science Conference 1988.

18. I. Hulthage, "Design of Specification for Knowledge Based System for Materials Design", report of Contract No. DAAL03-86-D-0001, 3 June 1988

19. J. Fiksel and F. Hayes-Roth, IEEE Expert 4, 16-23 (1989)

20. Daniel G. Bobrow, Ed.,"Qualitative reasoning about physical systems", MIT Press, Cambridge, MA 1985

21. L.E. Widman,Y.B. Lee, Y.H. Pao, "Towards the Diagnosis of Medical Causal Models by Semiquantitative Reasoning,",P.L. Miller,Ed.,Topics in Medical Artificial Intelligence, Springer Verlag, N.Y.,(1987)

22. e.g., ASTM Committee E-49 on Computerization of Materials Property Data; the Materials Properties Data Program of NIST Office of Standard Reference Data; MIST, Materials Information for Science and Technology at the Lawrence Berkeley Laboratory, supported by U.S. Dept. of Energy.

23. Proceedings of the June 1987 Workshop on New Issues on Materials Data-bases, Stanford University, G. Wiederhold, Editor.

24. AIMS project, Artificially Intelligent Materials Selection, a joint effort among the Library Research Corporation of America, the BDM Corporation, and George Mason University in cooperation with the National Materials Properties Data Network, Inc.

25. E. Hornbogen,"On the Microstructure of Alloys", Acta metall., 32,615-627(1984)

26. J. Hobbs,"Granularity",Proc. 9th Int. Joint Conf. on Artificial Intelligence, Vol.1,pp.432-435

MANUFACTURING CONTROL THROUGH THE DEVELOPMENT

OF A SYSTEM FOR REAL TIME MATERIALS PROPERTIES MONITORING

K. Kozaczek, C.O. Ruud, J.C. Conway, Jr., C.J. Yu
The Pennsylvania State University
University Park, Pa. 16802

ABSTRACT

The goal of this investigation is to develop and demonstrate the feasibility of a system for in-process interrogation of copper alloy strip. The system is to continuously provide a characterization signature to be used to determine the instantaneous properties and forming characteristics of the metal. The need for in-process characterization is based upon the broad scatter in properties and forming characteristics which exists in commercial wrought metal products, and the loss of productivity and product quality caused by that scatter. The investigative approach uses a unique combination of three concerted nonintrusive techniques, e.g., eddy current, ultrasound, and x-ray diffraction, all three of which individually have seen limited application in the real-time frames necessary. This paper will illustrate the correlations developed between nondestructive characterization parameters and the mechanical and forming properties of the copper base alloys.

Expert System Applications in
Materials Processing and Manufacturing
Edited by M.Y. Demeri
The Minerals, Metals & Materials Society, 1989

INTRODUCTION

Producing consistent results of the desired quality from manufacturing processes which involve plastic deformation of the material is often frustrated by the anisotropic properties of the feed stock. The material properties essential for the proper results from a manufacturing process differ from one feed stock supplier to another or within a feed stock supplied by a particular vendor. Furthermore, the properties often change continuously from one end to the other of a single feed stock reel. For example, the Metals Handbook [1] shows a 10 to 30 percent variation in tensile and yield strength as well as ductility of beryllium copper, an important electrical switch and connector alloy.

The necessity of adjusting the forming parameters to variations in the feed stock properties as well as the shutdowns in production and poor quality of the final product causes costly losses to industry. The uncertainty in mechanical properties results in the worst case design approach, which, as a consequence impeeds the optimum use of materials and design capabilities.

The sampling of feed stock in incoming inspection can not resolve all the forementioned problems. The scope of incoming inspection is limited to piece meal measurements of mechanical properties by means of time-consuming destructive techniques. Under many circumstances in-process characterization of forming properties is highly desirable and would be the base for development of expert systems. The research necessary for the development of such systems would include:

- o The understanding of the interrelationships between the micro and macroscopic character of the material and its mechanical properties essential for a particular forming process.
- o Demonstration of the feasibility of rapid on-line, non-destructive characterization of the relevant micro and macroscopic character of the material.

The forming properties by and large depend on elastic modulus, plasticity (ductility), yield strength, shear strength, Poisson's ratio and hardness. The anisotropy parameter (Langford's coefficient) appears to be the most important parameter for determining drawability in deep drawing while the strain hardening exponent and the strain ratio sensitivity are the most important parameters in stretch forming operations [2]. Unfortunately, for the most part, these properties are defined by destructive mechanical tests. An exact relationship between mechanical properties and forming characteristics are in most cases not available either. The most adequate approach to prediction of material behavior during forming is to monitor microstructural characteristics. The characteristic parameters of the microstructure can be employed for the determination of mechanical properties by means of development of empirical equations. For example, yield strength can be approximated by the equation:

$$Y.S. = \sigma_y = \sigma_o + K_y d^{-\frac{1}{2}} + \sigma_{CW} - \sigma_R + \Sigma\, a_i P_i + \Sigma\, b_i C_i$$

$$+ \sigma_t + K_{SC} S^n \,,$$

where σ_o = constant, K_y = grain size constant, d = grain diameter, σ_R = residual stress, a_i are the constants affecting the phase composition effect on yield strength, P_i is the volume fraction of each major phase (i), b_i are the constants affecting the element and composition effect on yield strength, C_i is the weight function increase (or decrease) due to texture, K_{sc} = strength coefficient, S is the plastic strain imposed during fabrication, and n = strain hardening exponent [3]. The formability is generally understood as the limiting strain to which material can be deformed prior to fracture. The formability of sheet metal depends on the forming strain limit and the uniformity of the strain distribution. The strain distribution depends on the unique characteristics of the material such as texture, residual stress, dislocation density, precipitates, chemical composition, changes in thickness, surface roughness etc. There are two basic concerns in the development of a system for nondestructive, real-time characterization of metals. First, the feasibility of rapid, nondestructive measurement of essential microstructural parameters has to be demonstrated. Secondly, through mechanical testing, one has to relate the microstructure to formability defined as the materials ability to sustain deformation tests simulating the forming process.

<u>INVESTIGATIVE APPROACH</u>

The isolation and independent measurement of all the characteristics of the metal that affect the forming process are not yet attainable [4]. However, specific mechanical properties can be represented by a "signature" function of the metals characteristics. A number of "signatures" representing properties of materials and dependent upon several nonintrusively determined parameters are being developed in an investigation at Penn State University.
Three nonintrusive techniques are being used to provide a composite signature. The techniques that have been selected are measurement of acoustic velocity and attenuation of ultrasonic Lamb waves, x-ray diffraction measurements using position sensitive detectors, and conductivity measurement through eddy current inducement.
The investigative approach was to use a unique combination of three concerted techniques which would be sensitive to different characteristics in complimentary ways. The teamed signature would provide a sensitive indicator for the thermo-mechanical condition of metal. The multivariable semi-empirical equations relating mechanical and forming properties to teamed signatures will be developed as an effective tool for process control.

<u>SPECIMENS</u>

C110 copper specimens were supplied in strip form 12x6x0.01". There were two sets with different grain size 5 Ìm and 19 Ìm. Each set consisted of four samples, one sheet with "random texture", and 15%, 45% and 75% cold rolled. The thermomechanical history of the samples was also controlled. Standard mechanical and metallography tests were conducted on copper samples providing information about yield strength, elongation, Young's modulus, surface roughness and microstructure.

<u>ULTRASONIC VELOCITY MEASUREMENT</u>

The ultrasonic velocity measurements offer a practical way to extract quantitative information concerning crystallographic orientation, dislocation density, and size of grains in materials. Propagation velocities are determined by volume fractions of different grain orientations (types of crystallographic texture) and the directional constants of the grains. Numerous papers have been published regarding the dependence of acoustic wave propagation on texture. The theoretical calculations of the propagation velocity in a particular direction are usually based upon the assumption that Voigt, Reuss, or Hill models can be applied. Such an approach does not account for other parameters affecting propagation velocity. The measurements of velocity propagation of leaky Lamb waves in thin copper sheet conducted in this investigation have demonstrated that the grain size and stress fields due to dislocations have an influence on propagation velocity of ultrasound. Figures 1 and 2 represent the propagation velocity of leaky Lamb waves measured in several directions in specimens with different grain sizes and deformation levels. The increase of the deformation level from 15% to 75% cold rolling in the copper causes a transition of texture from (112) <111> (copper type) to (110) <112> (brass type). Also, it was determined that the larger grain size corresponded to higher acoustic velocities. The unexpected shape of the velocity plots for 15% reduction indicates the strong dependance of the propagation velocity on the stress field due to the dislocation tangles. For this deformation level (15%) the dislocation movement in (111) plane is the dominant deformation mode. Experimental results have demonstrated that grain size and cold work are explicit parameters which determine the dislocation density and preferred orientation of grains in cold rolled copper. The measurement of S_0. mode Lamb wave velocity allows for the monitoring of texture induced anisotropy. The use of the normalized velocity factor [5] provides for the separate grain size and cold work effects. The results of this investigation to date indicates that the ultrasonic velocity measurement technique shows promise for incorporation in an on-line, in-process, nondestructive material property measurement system.

ADVANCED X-RAY DIFFRACTION TECHNIQUE

An advanced XRD technique suitable for monitoring the nature of the texture in thin copper sheet and providing supplementary information about the cold work and residual stress in the copper has recently been presented by the authors [6]. The XRD system with a position sensitive scintillation detector enables the measurement of intensity of the diffraction peaks, peak shape, and peak shift in time frames of a few seconds. This technique permits the monitoring of the texture condition of the material. Figure 3 illustrates the measured change in grain orientation as related to the level of deformation of cold rolled copper.
The XRD technique can also be used for the measurement of microstresses (stresses due to strain incompatibilities between grains or subgrains) in various crystallographic directions. The measured spatial and directional distribution of microstresses can be used for monitoring the deformation mechanism. Figure 4 shows the predicted lack of measured microstresses in the plane of easy slip where the presence of strain incompatibilities is very unlikely.
The XRD technique was also applied to the measurement of the macrodeformation in selected crystallographic directions in the material. The measurement of changes in the spacing between crystallographic planes provides for the determinations of residual stress tensors [7]. The feasibility of the measurement of residual stresses in the thin copper strip moving at 375 ft/min has been demonstrated at Penn State recently [8]. Both, the state of residual micro and macro stresses and the texture strongly affect the formability of polycrystalline material.
The XRD measurements have been shown to provide a means of

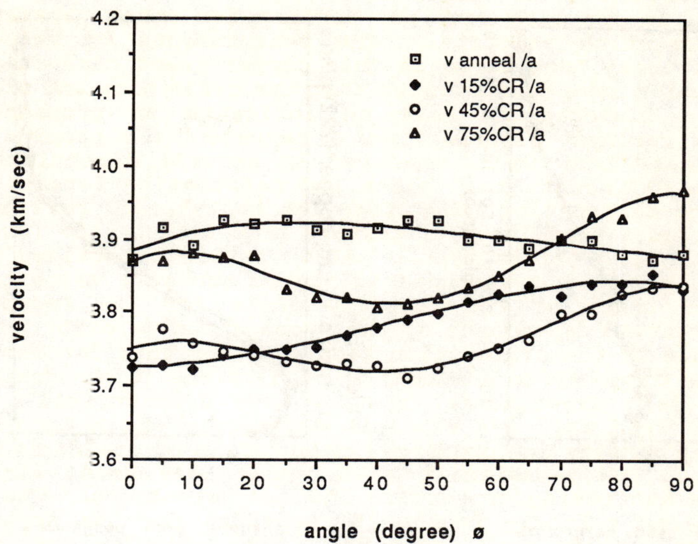

Fig. 1. Plots of ultrasonic velocity versus ϕ angle (measured in rolling plane from rolling direction) for different deformation levels, grain size 5 μm in annealed state.

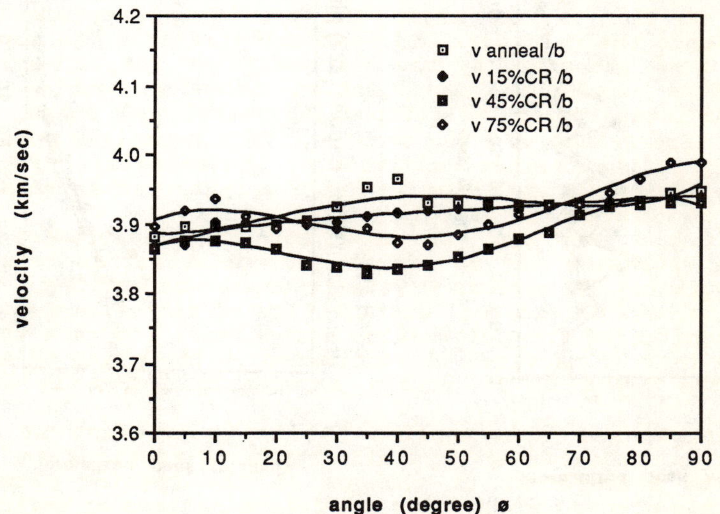

Fig. 2. Plots of ultrasonic velocity versus ϕ angle (measured in rolling plane from rolling direction) for different deformation levels, grain size 20 μm in annealed state.

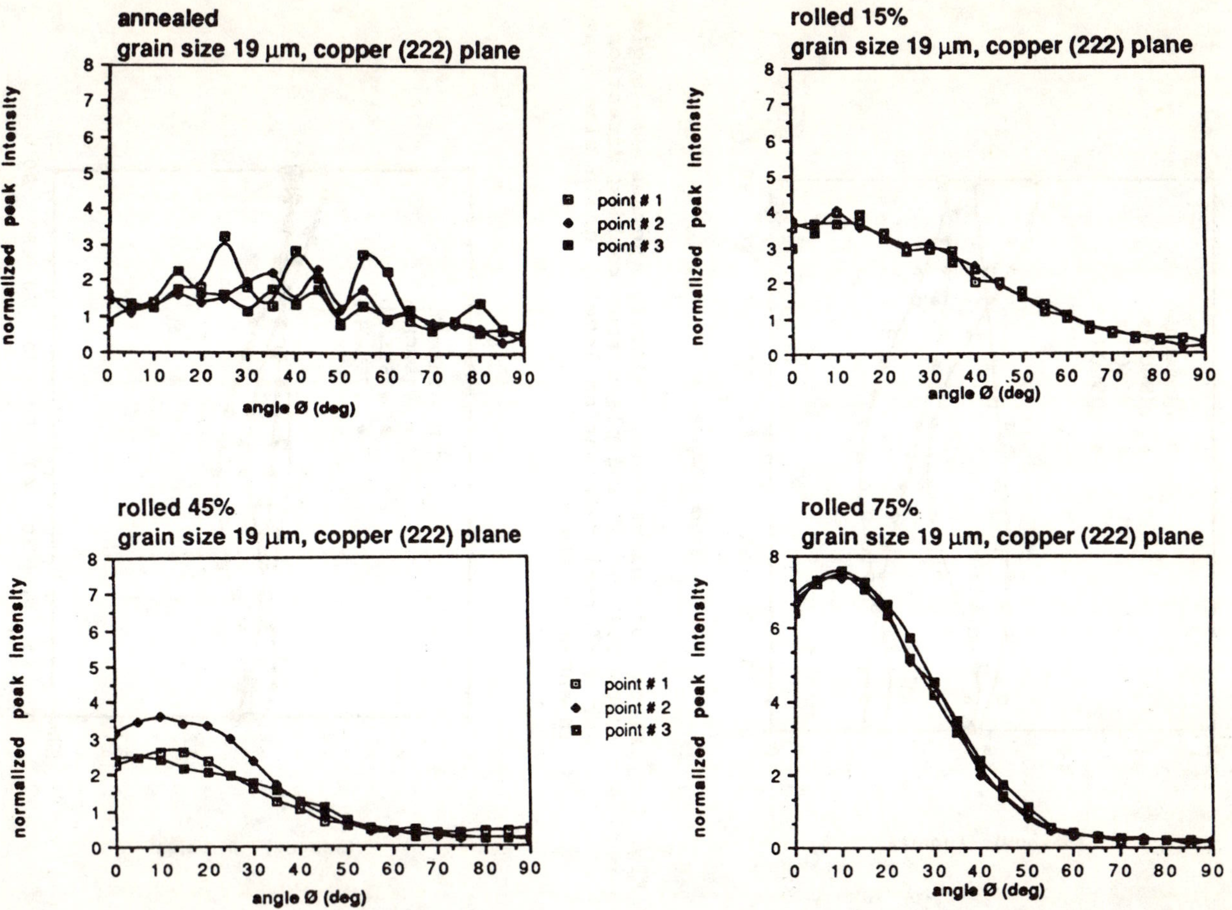

Fig. 3. Changes in grain orientation for different deformation levels represented by diffraction peak intensity measured in polar coordinates; ϕ-angle measured in rolling plane from rolling direction.

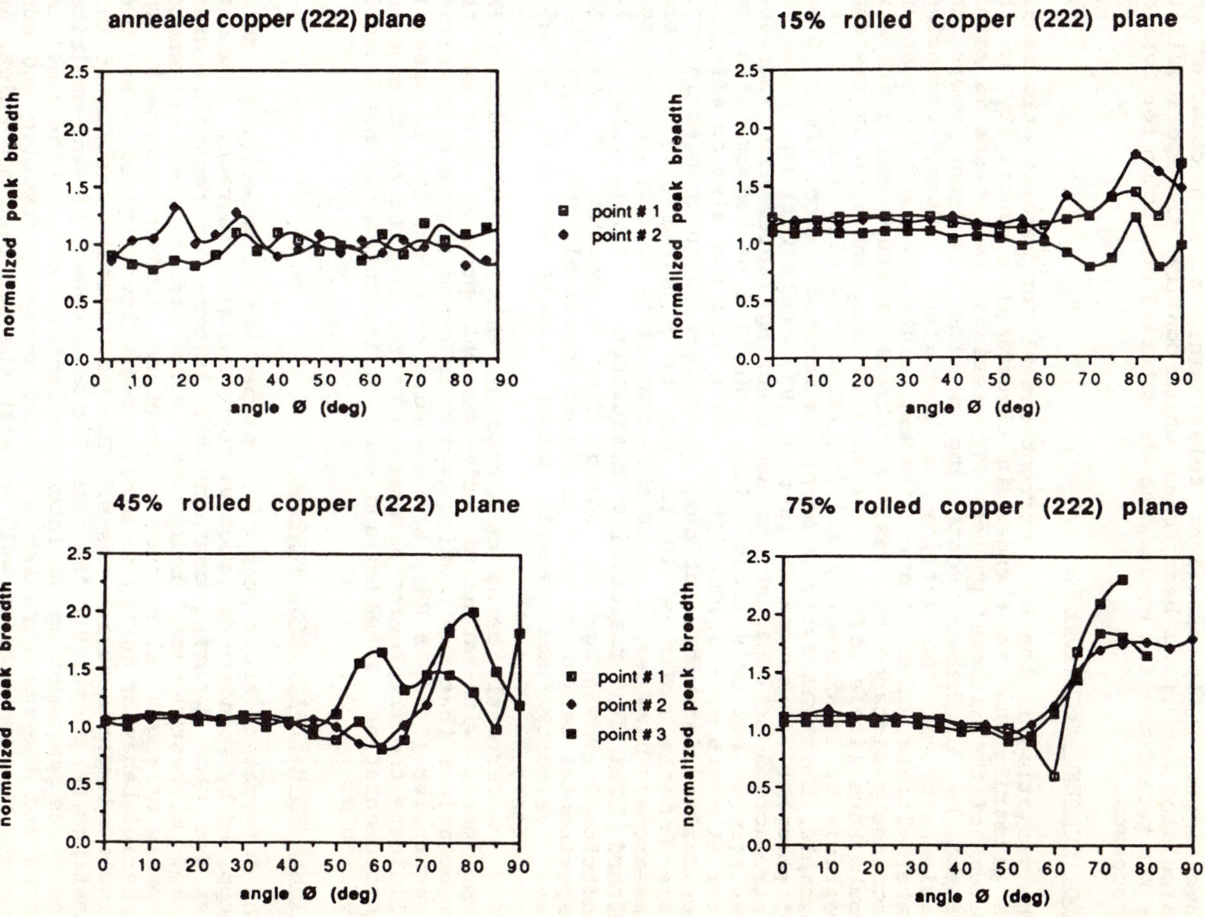

Fig. 4. Microstress represented by diffraction peak breadth on (222) family of crystallographic planes in copper; ϕ-angle measured in rolling plane from rolling direction.

monitoring texture on thin cold rolled copper sheet. The correlation between certain x-ray diffraction parameters and observed earing values for aluminum alloy has also been investigated. An excellent relationship between observed and calculated earing has been found, which confirms the applicability of XRD techniques for monitoring the material condition for forming processes.

<u>EDDY CURRENT TECHNIQUE</u>

The variation in the eddy current signal for paramagnetic metals is primarily due to changes in electrical conductivity in the volume of the material probed. The residual resistance is mainly caused by incoherent scattering of the electrons where the periodicity of the lattice is disturbed. Thus, vacancies, interstitials, coherent precipitates, solute atoms and dislocations increase resistance but residual stress, texture and grain size would have little effect. The most drastic increase in residual resistivity is caused by foreign atoms in solid solution. For example, the resistivity of copper is increased about 33% for 0.06% of phosphor introduced in lattice [9]. Cold working, i.e., the introduction of dislocations and stacking faults also increases resistivity but only about 20%for every 10% of reduction in area (for Al - 0.6 Ni - 0.4Sb) [10]. Grain size has also an effect on residual resistivity, but only relatively large changes in grain size (of the order of 10 İm) can be measured [11]. Several researchers have demonstrated the feasibility of eddy current decay method ECDM for resistivity measurement. [10,11,12,13]. The additional advantage of using this technique is that the experimental set up for ECDM is almost identical with that for Lamb wave velocity measurement using electromagnetic acoustic transducers.
ECDM is currently being developed in this investigation. The experiments with standard eddy current probes confirmed the expectations that for testing very thin brass strips high testing frequencies (10 - 15 MHz) are needed. Additionally, in order to eliminate the irrelevant parameters of testing the multifrequency instrumentation is needed and such an instrument is not readily available.

<u>SIMULATED MANUFACTURING TESTING</u>

The information about mechanical properties of material are to be supplied by standard destructive tests and metallography. The data set for each material contains the information regarding grain size, microstructure, thickness, 0.2% offset yield strength, tensile strength, elongation of fracture, surface roughness. The material behavior during forming is being investigated by means of simulated manufacturing tests. The quantitative results of this testing will constitute the basis for development of semi-empirical equations representing the interrelationships between the simulated tests and the nonintrusive acoustic, x-ray diffraction and eddy current measurements as well as with the standard mechanical and metallurgical tests.
The selection of quantitative, dynamic mechanical test simulating forming processes depends on each process. For example, one manufacturer of electronic switches has developed a test die set for predicting bend formability. This one-out progressive die can blank a variable width finger of material and bend it over a variable inside radius through a 90° bend. The results from this test correlate well with production die behavior.

SUMMARY

The proceeding test reveals the rationale of how empirical team
signatures will be developed to manifest the mechanical properties
and stamping and forming characteristics of the copper alloy strip.
The results which have been obtained to date, indicate that
acoustic and XRD techniques, can provide the essential information
about material microstructure in time frames suitable for on-line
applications. This information together with data obtained from
eddy current tests and simulated manufacturing tests will create
the foundation for development of a teamed sensor (fused sensor),
expert system for intelligent process monitoring and control.

REFERENCES

1.   ASM, Metals Handbook, 9th Ed. 1979.
2.   Kwang, Y., Morris, J.G, "The Effect of Structure on the
Mechanical Behavior and Stretch Formability of Constitutionally
Dynamic 3000 Series Aluminum Alloys," Materials Science and
Engineering, 77 (1986), 59-74
3.   Guy, A.G., Hren, J.J., Elements of Physical Metallurgy, 3rd
Ed., Addison-Wesley, 1974,137.
4.   Ruud, C.O. Vikram, C., "Morphological Investigation of
Fundamental Methods of In-Process Materials Evaluation for
Productivity Improvement," Proceedings of the 12th Conf. on
Prod. R. and D., Nat. Science Foundation, Pub. by Soc. of Manf.
Engr., ISME Dr., Dearborn, MI, May 1985, 345-352.
5.   Yu, C.T., Conway, Jr., T.C., Ruud, C.O., Kozaczek, K.,
"Ultrasonic Monitoring of Textures in Cold-Rolled Copper
Sheets," to be published in Review of Progress in Quantitative
Nondestructive Evaluation, 1989, Pub. by Thomson, D.O.,
Chimenti, D.E.
6.   Kozaczek, K., Ruud, C.O., Conway, Jr., J.C., Yu, C.J., "The
Representation of Texture in cold-rolled Copper Sheet by an
Advanced X-Ray Diffraction Technique", to be published in
Reviews of Progress in Quantitative Nondestructive Evaluation,
1989, pub. by Thomson, D.O., Chimenti, D.E.
7.   Ruud, C.O., Chen, P.C., "Application of an Advanced XRD
instrument for Surface Stress-Tensor Measurements on Steel
Sheets", Experimental Mechanics, Vol. 25, No. 3, September
1985, 245-250.
8.   Ruud, C.O., Private Communication, May 1989.
9.   ASTN Nondestructive Testing Handbook, ed. by R.C. McMaster,
Vol. 2, 1959, pp. 4212-4213.
10. Hartwig, K.T., "An Eddy Current Decay Technique for Low-
Temperature Resistivity Measurements", Eddy-Current
Characterization of Materials and Structures, ASTM Special
Technical Publication 722, ed. Birnbaum, G., Free, G., 1979,
157-172.
11.   Drew, R.A.L., Muir, W.B., Williams, W.M., "Differential
Measurement for Monitoring Annealing," Metallurgical
Transactions A, Vol. 14A, February 1983, 175-182.
12. Bean, C.P., Deblois, R.W., Nesbit, L.B., "Eddy-Current Method
for Measuring the Resistivity of Metals", Journal of Appl.
Phys., Vol. 30, No. 12, Dec. 1959, 1976-1980.
13. LePage, Bernalte, A., Lindholm, D.A., "Analysis of Resistivity
Measurements by the Eddy Current Decay Method", The Review of
Scientific Instruments, Vol. 39, No. 7, July 1968, 1019-1076.

EXPERT SYSTEMS IN FORMING PROCESSES

M. Y. Demeri

Manufacturing Systems Department
Research Staff
Ford Motor Company
Dearborn, MI 48121

Abstract

Today's highly competitive market mandates on the manufacturers to make major changes in their traditional product development cycle from concept through production. There is a strong need to reduce cost, shorten development cycle and improve product quality. To achieve this goal, the traditional product development process, which is based on the engineers's reliance on his experience to build and test new products, needs to be improved. The time and expense for product development increases as the number of design trials increase. The traditional product development process works fine for minor modifications but becomes highly inefficient for new designs and products. Engineers need new tools to evaluate alternative designs and manufacturing processes before prototype building and testing. This paper addresses the potential integration of Expert Systems with Process Modeling and Simulation. It also reviews the benefits that such an integration would extend to applications in the area of materials forming processes.

Expert System Applications in
Materials Processing and Manufacturing
Edited by M.Y. Demeri
The Minerals, Metals & Materials Society, 1989

Introduction

The traditional engineering process, shown in Fig. 1, is based on the engineer's reliance on his past experience and talent to build and test new products in a time consuming and highly expensive iterative process. The time and expense increases as the number of design trials increase. For minor modifications, the traditional process is suitable but for new designs and products it becomes highly inefficient. Prior experience is of little help as designs, materials and new processes change dramatically.

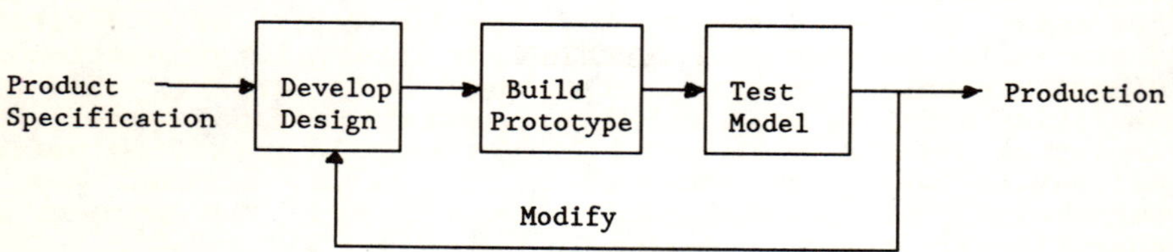

Fig.1 Traditional Approach to Manufacturing.

To meet the challenge, new methods and tools have to be developed and used. Engineers need tools to evaluate alternative designs and manufacturing processes before prototype building and testing. Software tools are more efficient and, in the long run, cost less than hardware tools. Computer Aided Design tools are available for the design engineer and process simulation tools are available for the manufacturing engineer.

Computer simulation refers to computing in support for model building and model implementation. Simulation techniques utilizing geometric, analytical and finite element methods have been developed to predict, as in the case of forming, the plastic flow of materials during the forming process. Process simulation helps the stamping and product engineers to evaluate the manufacturing feasibility of parts at the design stage. Through simulation, alternate designs can be explored and trade-offs can be evaluated to arrive at the best design with the right material at the lowest cost and the shortest lead time.

<u>**Computer Simulation and Modeling**</u>

Simulation models are representations of a system developed with sufficient detail to enable the analyst to make decisions about the behavior of the system. In this case, the system is the material forming process and the model is the set of assumptions about how the system works. These assumptions constitute a model which usually takes the form of mathematical relationships. If the relationships describing the model are fairly simple, analytical solutions can be used to obtain exact information about the system. Many systems are too complicated to be evaluated by analytical means and can only be evaluated by using simulation techniques. A number of models may exist to represent the behavior of a given system. A simulation model of the mathematical type can be static (time-independent), dynamic (time-dependent), deterministic (contains no random variables), stochastic (contains one or more random variables), discrete (state variables change at certain time intervals) or continuous (state variables change continuously with time). If the simulation model is not a valid representation of the system, the simulation results will be of little help in providing useful information about the behavior of the actual system. Once a reasonably accurate model has been selected, the process of generating its behavior begins.

<u>**Expert Systems**</u>

Expert Systems are emerging now as a potential tool to increase the efficiency and productivity of manufacturing processes. An Expert System is an intelligent computer program that uses knowledge and inference procedures to solve problems that usually require human experts to solve. Knowledge-based programs contain knowledge from experienced and talented people in certain fields. The programs use this knowledge to emulate the inferential reasoning of the experts to find solutions to specific problems . Knowledge systems contain three components: a knowledge base, a reasoning engine and a control mechanism.

Knowledge-based information is acquired by interviewing experts in the field and representing the information in a special code. There are four methods of representing knowledge:

1. Semantic Networks (uses predicates and attributes to represent objects and show relationship between them).

2. Rules (if-then-else format).

3. Frames (symbolic knowledge representation structure that holds information in memory areas that contain facts, rules, actions, etc.).

4. Logic (provides means for performing deduction).

The choice of the method for knowledge representation depends on the application intended and the best way to retrieve and deduce the information from the knowledge base. The reasoning engine uses two strategies: forward chaining (program finds a solution from a set of facts and rules) or backward chaining (program assumes a solution is true, then it chains backwards through the facts and rules to find evidence to support its assumption). The function of the control mechanism in a knowledge system is to direct the search for a solution. Details on the subject of knowledge representation are beyond the scope of this paper. Also, there will be no reference to the hardware needed to handle such programs.

Expert Systems and Computer Simulation

Expert systems can assist in the simulation process by choosing the most appropriate model, from a library of all possible and acceptable models, and by selecting the best method of solution, from a library of available and efficient solution methods, to reach a well defined goal. An expert simulation system would be designed to automatically find the best model and the appropriate solution method to fulfill the specified goal. According to O'keefe (1), expert systems and simulation can be integrated if results from prior simulations are used as input data to an induction algorithm which then deduces production rules. Cunningham (2) believes that a model can be used as the inference component of an expert system. Expert systems are knowledge-based and, unlike data-based systems, are more general because they contain assumptions about the real system which go beyond available data and thus enable prediction and forecasting which are essential characteristics of any useful simulation. Process simulation usually includes the following components:

1. Mathematical formulation of the process (welding, forming).

2. Data base (material, geometric, design guidelines).

3. Part design (shape, contours, radii).

4. Process Design (forming sequence, equipment).

Process and part design are usually based on previous knowledge, experience and artistic touch of the process and design engineers. This art and experience can be captured by knowledge engineers and coded into a knowledge-based expert system for utilization in part production. An expert simulation system would work according to the following sequence:

A. The expert system recommends a simulation based on the
   information residing in its knowledge base.

B. The simulation is run and results obtained.

C. Results are evaluated and problems identified.

D. A diagnostic expert system suggests remedial actions
   to correct problems.

E. Iteration continues until problems are eliminated.

According to Maddux and Jain (3), expert systems can serve two functions: One is to assist in defining the initial processing conditions for the simulation process and the other is to diagnose and recommend modifications to correct the problems. A flow chart of an Expert Simulation System of this type would look like this:

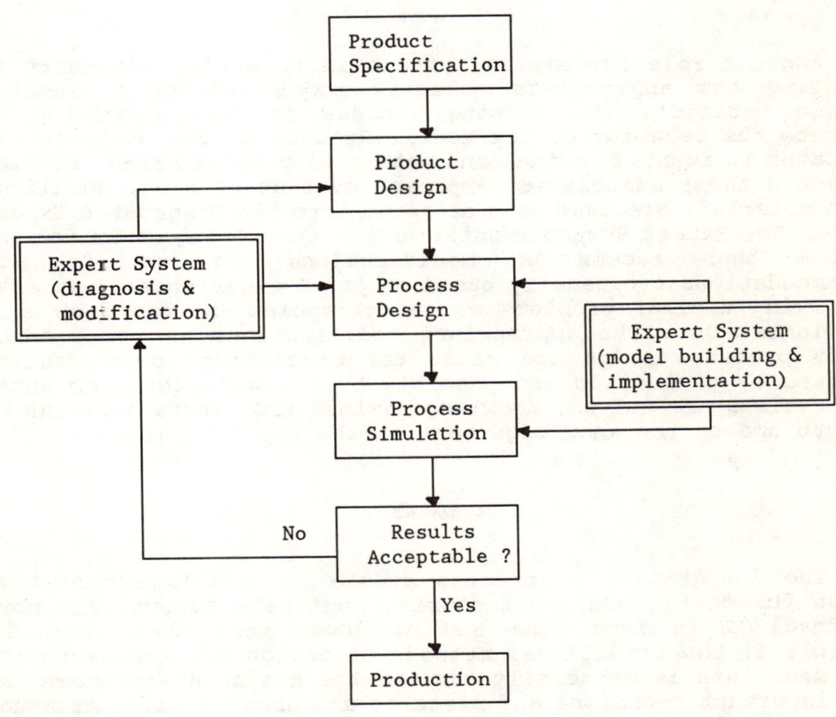

Fig.2 Flow Chart of an Expert Simulation & Diagnostic System

The integration of Expert Systems and Computer Simulation in the forming process is an emerging technology. The importance of an Expert System is emphasized by the fact that part geometry, process development and tool design are very much based on prior experience and rules of thumb. This experience must be captured and coupled with model simulation to solve material forming problems.

In its initial guess role as input to help in developing part shape and defining process parameters, an expert system may define contours, radii, preform shapes, tip angles and draw bead configurations based on key design features and guidelines stored in the knowledge base. The design process iterates until a satisfactory design is obtained. Once the part geometry is defined, the processing approach is chosen and the process parameters (die, tooling, fixtures) are picked from a number of alternative variables stored in the knowledge base of the expert system.

Another role for expert systems is to assist engineers in selecting the appropriate material model (elastic, plastic, dynamic, static). The forming process is then simulated to generate the behavior of the model. Results of the analysis are evaluated to identify actual and potential problem areas. Problem areas and their associated symptoms (over-stretching, buckling, loose material) are then used as input into the Diagnostic Expert System. The Expert System identifies the underlying cause for the symptom and recommends modifications to rectify it. Recommendations are usually based on prior experience of experts in solving similar problems or on extrapolations based on such experience. Once the appropriate modifications in the product and/or process design are made, the simulation is run again, followed by evaluation and possibly by further iteration until all problems are solved. Knowledge gained from iterations can be used to update the knowledge base in the expert system.

Summary

The integration of process simulation and Expert Systems allows the design and manufacturing engineers to evaluate more alternatives in less time and at lower cost than would be possible if the traditional methods of design and manufacturing are used. This is especially true at the design stage where the most important decisions and mistakes are usually made. Although some commercial simulation and knowledge systems software appear on the market with increasing frequency, custom-building of highly specialized software will continue to be the trend, at least, for the near future.

<u>References</u>

1. R. O'Keefe, "Simulation and Expert Systems - A taxonomy and some examples", Simulation, 46 (1), 1968.

2. J. Cunningham, "Comprehension by Model-building as a Basis for an Expert System", Expert Systems 85, M. Merry ed. (Cambridge University Press, 1985).

3. K. Maddux and S. Jain, Artificial Intelligence Applications in Materials Science, AIME/ASM Conference Proceedings, Orlando, 1986, TMS Publication.

AN EXPERT DECISION SUPPORT SYSTEM

FOR DEFECT PREDICTION IN FORMING PROCESSES[1]

Namita Goyal[2] and H.A. Kuhn[3]

Abstract

An overriding concern in today's manufacturing industry is to produce quality parts economically without compromising mechanical and physical properties. In precision forging processes for example, manufacturing defect-free parts is of prime importance. This requires systematic determination of potential fracture regions and specification of measures to avoid fracture. The approach taken for automated application of a fracture criterion is to build an integrated process design framework based on a set of advanced software tools which perform reasoning with hybrid data (both symbolic and numeric). An Artificial-Intelligence based decision-support system is implemented using KEE, an expert system shell, on a SUN workstation. The Finite Element Method is used selectively for simulation and validation of the potential fracture regions identified by the decision-support system. The design and control of the system is influenced by the selection of part and preform geometry, material and other processing conditions (eg., workpiece and die temperatures, lubrication).

Expert System Applications in
Materials Processing and Manufacturing
Edited by M.Y. Demeri
The Minerals, Metals & Materials Society, 1989

---

[1]This work was carried out while both authors were at the University of Pittsburgh

[2]Presently working at Carnegie Group, Inc.

[3]Presently Vice President, Metalworking Technology, Inc.

<u>Introduction</u>

Advances in technology bring about improvements in processes, materials and methods of process design and control. The Powder forging process combines the unmatched advantages of powder metallurgy along with high strength of forged parts. Powder forged components exhibit unique properties in terms of economy and quality which cannot be obtained by any other metal forming process. Some of these benefits are:

- Precise control which leads to uniformity and optimum performance characteristics

- Custom made compositions

- Unusual physical properties attainable eg., self lubricant, self damping characteristics

- Reduced machining and finishing

- Little or no material loss

- Reproducibility

The conventional approach to design of powder forging processes employs purely empirical trial-and-error procedures and rules-of-thumb on production equipment. The shortcomings of this approach are that even though success may be achieved ultimately, labor costs, tooling development costs, and lead times are excessive.

With the advent of analytical techniques like the Finite Element Method (FEM), localized stress and strain distributions may be provided for metal flow during die filling. Finite Element Modeling, however, is complex in both operation and interpretation of numerical output, even in graphic form. It requires skilled engineers to set up an accurate model which involves meshing, specification of material characteristics and definition of boundary conditions. Also, interpretation of the simulations obtained requires a solid foundation in mechanical metallurgy. Coupled with these problems, an expensive and time consuming simulation is obtained for the entire part, which may not be necessary.

An alternative approach is to employ *computer-aided predictive schemes* which take advantage of common sense and experiential knowledge based on past successes to explore process design and successfully implement a design strategy. This approach not only limits the search to a few alternatives but also avoids the reliance on complex analysis capabilities and computer facilities needed for using analytical tools. Thus, sophisticated computer based tools have been developed to take these operations from experiential and heuristic methods into analytical models which are finding increased acceptance by manufacturers worldwide.

<u>Knowledge-Based Synthesis</u>

The design and control of powder forging processes depend on an understanding of the characteristics of the workpiece material, conditions at the tool/workpiece interface, mechanics of plastic deformation (metal flow), equipment used and finished product requirements. These factors influence the selection of tool geometry and material as well as processing conditions (e.g. workpiece and die temperatures, lubrication). Artificial Intelligence (AI) methodologies provide computer based tools which can be combined with process, material and manufacturability knowledge to devise a control strategy for fracture prediction and avoidance. The primary objective of this work is to model the human decision-making process of designing defect-free parts. This involves a rational approach to understanding the sources of fracture and determining remedial procedures. It is conceptualized to build an integrated design framework using interactive procedures. The major components of the system are:

- A CAD description of the part to be formed. It displays a metrical representation of the shape including nodal points defined by X,Y coordinates, radii and other geometric details.

- An AI based decision support system which determines the possibility and probability of each of the various types of fracture occuring in different regions of the part based on heuristics, experiential and

28

laboratory knowledge of the process. It also has a database of powder characteristics, material properties and manufacturing process capabilities. Classification and reasoning of defects is done using a rule-system approach. The system would also suggest design changes in the preform geometry, tooling parameters or process conditions as necessary. This is implemented using KEE, an expert system shell running on a SUN 4/286 workstation.

- FEM is used selectively for simulation and validation of the potential fracture regions identified by the decision-support system.

The work presented here is a step toward complete and reliable automation of the design process. Our main focus in the present work is on the AI approach to representation and reasoning of various types of defects. AI techniques can be combined with process, material and manufacturability knowledge in order to devise a control strategy for fracture detection.

Fracture And Workability

Previous studies (15, 13) have shown that surface fracture strain combinations fit a locus having a slope of one-half for all materials, but the intercept at the ordinate depends on the ductility of the material. These results suggest a graphical representation of the interaction of material variables with process variables, Fig. 1. Material fracture limits are represented by straight lines with slope one-half, and the process is represented by strain paths in various potential fracture sites.

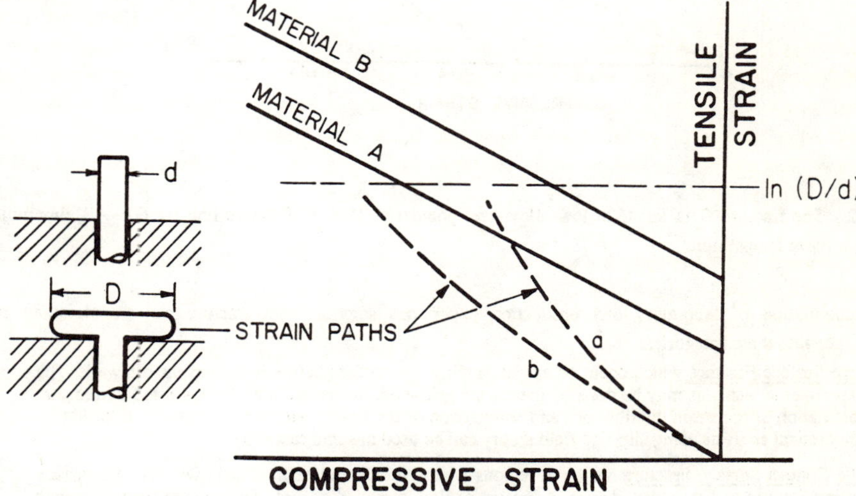

Figure 1 - Superposition of strain paths on forming limit lines (2).

As a simple illustration, consider the formation of a bolt head by upsetting the end of a rod. If it is intended to form a bolt head diameter D from a given rod of diameter d, the required circumferential tensile strain will be ln(D/d), as indicated by the horizontal dotted line. Consider that two different materials A and B have forming limit lines as shown by solid lines. Suppose strain path 'a' describes the progressing strain state for a particular friction condition. In order to reach the bolt head diameter, d, the forming limit for material A is crossed and fracture is likely to occur. Cracking can be avoided by using material which has the forming limit line B above the termination of the strain path 'a'. Alternatively, the process variables may be changed such that the strain path alters to 'b', and again the final bolt head diameter is reached before the forming limit line for material A is crossed. This can be accomplished in the case of bolt heading by improving lubrication, which increases the magnitude of the surface compressive strain and decreases the surface tensile strain. In other cases, the strain path can also be controlled by imposing changes in the preform and die geometries.

The deformation of aluminium powder compacts at room temperature can be used to model the metal flow and fracture locations in hot forging of steel powder, Fig. 2. This result provides the basis for laboratory studies of the workability of various generic shapes in forging (4). Complex forged parts can be subdivided into regions, each composed of one of these generic shapes. Knowledge about defects in these simpler shapes is applied to design the process for the complex forging.

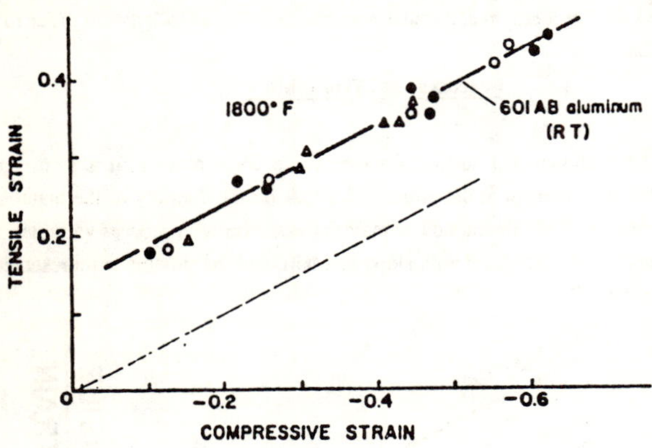

Figure 2 - The fracture Data for 4620 low alloy steel powder at 1800 F. Fracture line for 601 AB aluminium powder at room temperature.

The accumulation of laboratory and production experience suggests that fracture can be classified for convenience into three categories:

- <u>Free Surface Fracture</u> which occur on any expanding free surface before it reaches the die walls. The experimental analysis may be done by measuring grid mark deformation at the possible fracture site, calculation of the strain distributions and comparison of the strains with the fracture line as in Fig. 1. Theoretical analysis using slip-line field theory can be used in some cases.(5)

- <u>Die Contact Surface Fracture</u> occurs in regions where the material is in contact with the die surface during forging and a sudden change in friction or flow direction occurs. Die contact fracture occurs when there is a large gradient of frictional shear stress, rather than a uniform low or high friction condition.

- <u>Internal Fracture</u> usually occurs at the center of the workpiece and rarely penetrates to the surface; thus, it can remain undetected. Sometimes, internal fractures are closed up by subsequent deformations, but the resulting bond may not be sound metallurgically.

System Architecture

The fracture characteristics and workability data described previously have been formulated into a forging design support system. Details of implementation of the system is presented by describing a functional definition of each of the data items:

1. CAD system description of geometric primitives (i.e. lines, nodes, surfaces).

2. Database of powder characteristics, material properties and manufacturing process capabilities.

3. Frame based representation of the die cavity subdivided into regions, with manufacturing features as

one of the defining entities. Surfaces and nodes determe taxonomic relationships between objects. Both declarative and procedural information about the objects are represented.

4. Frame based representation of cofeatures, which constitute each of the manufacturing features.

5. Natural language-like interface for ease in communication between various components of the system.

6. Rule system for doing rule-based reasoning about defects. Rules relating to each particular subclass are bundled into a separate ruleclass. Traces are used for displaying goals and sub-goals in the deriving process.

7. User-friendly interface for viewing and modifying process and geometric conditions.

8. Modeling of different hypothetical situations to represent alternative strategies using context approach.

9. Graphical output from FEM analysis.

<u>**Representation and Reasoning**</u>

Both deterministic and heuristic knowledge is incorporated for building the knowledge base. Deterministic knowledge is obtained from design handbooks and previous experiments, while heuristic knowledge is gathered from experts in a series of interview sessions. The knowledge based system employs explicit knowledge structures (e.g. rules and facts). Facts are knowledge descriptions of the variables in the problem. Rules represent the knowledge of how to put facts together to generate a solution. An inference engine makes decisions within a domain by employing the problem solving methods and strategies neccessary to combine the rules and form a solution, Fig. 3.

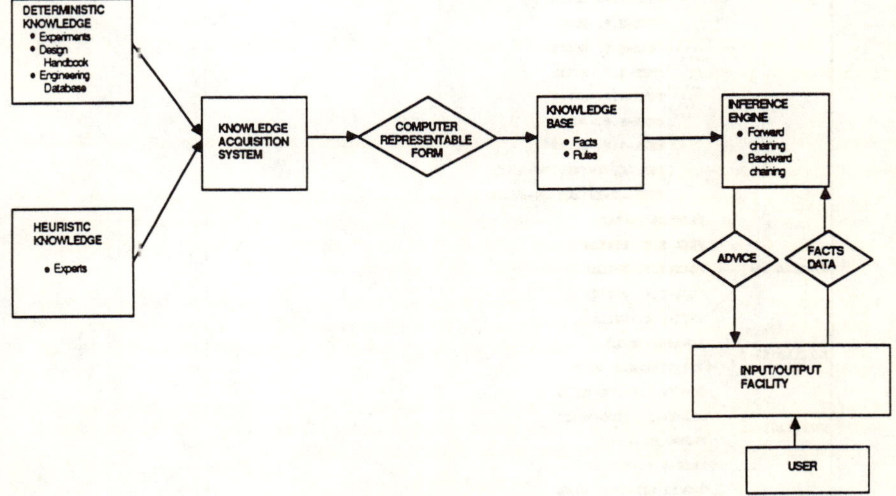

Figure 3 - Building blocks of a knowledge-based system.

The frame based representation of process, die and preform is very flexible, expressive and accessible for interpretation to the rest of the system. It is used to describe hierarchical relations for representation of objects and their attributes. The powder forging process has been subdivided such that each subsystem has a characteristic structural behaviour. The inherited information can be partially or completely modified as needed to help create a realistic model. (Fig. 4)

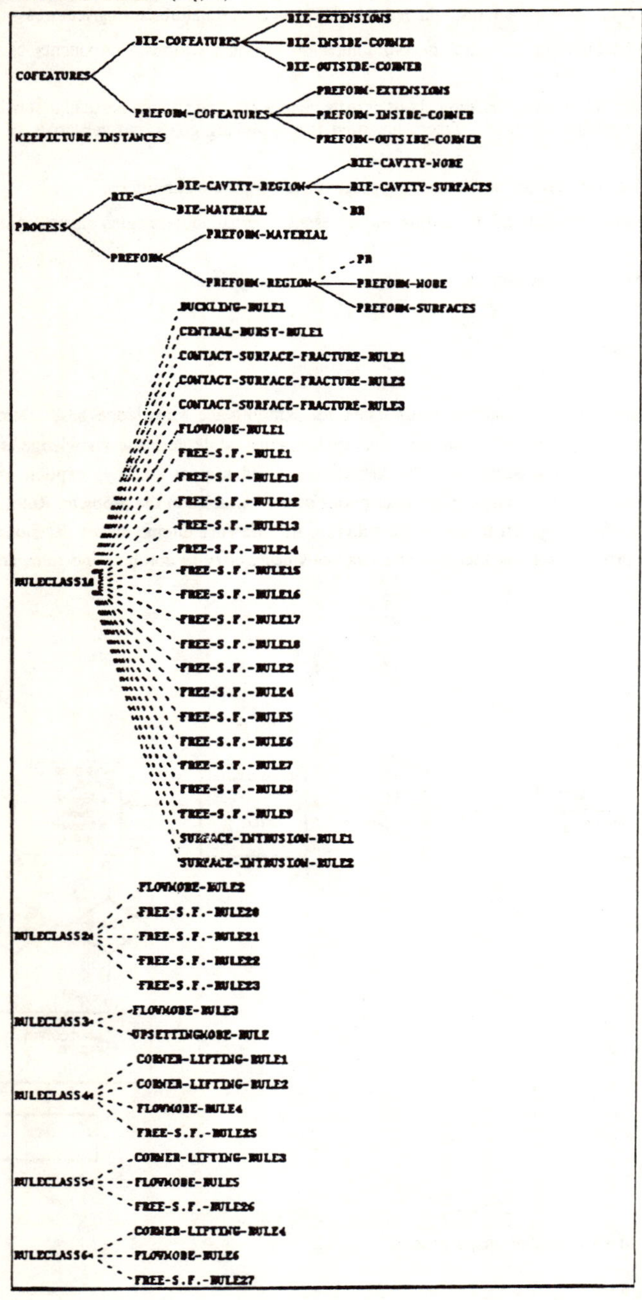

Figure 4 - A symbolic description of components of the knowledge base.

The attributes of each unit which is a data structure defined to represent an object are in terms of numeric, textual data and also procedural information As shown in Fig. 5, for example, the Preform-region frame has a number of attributes each specifying the range of qualitative values of physical parameters. Other examples are illustrated in Fig. 6 and Fig. 7.

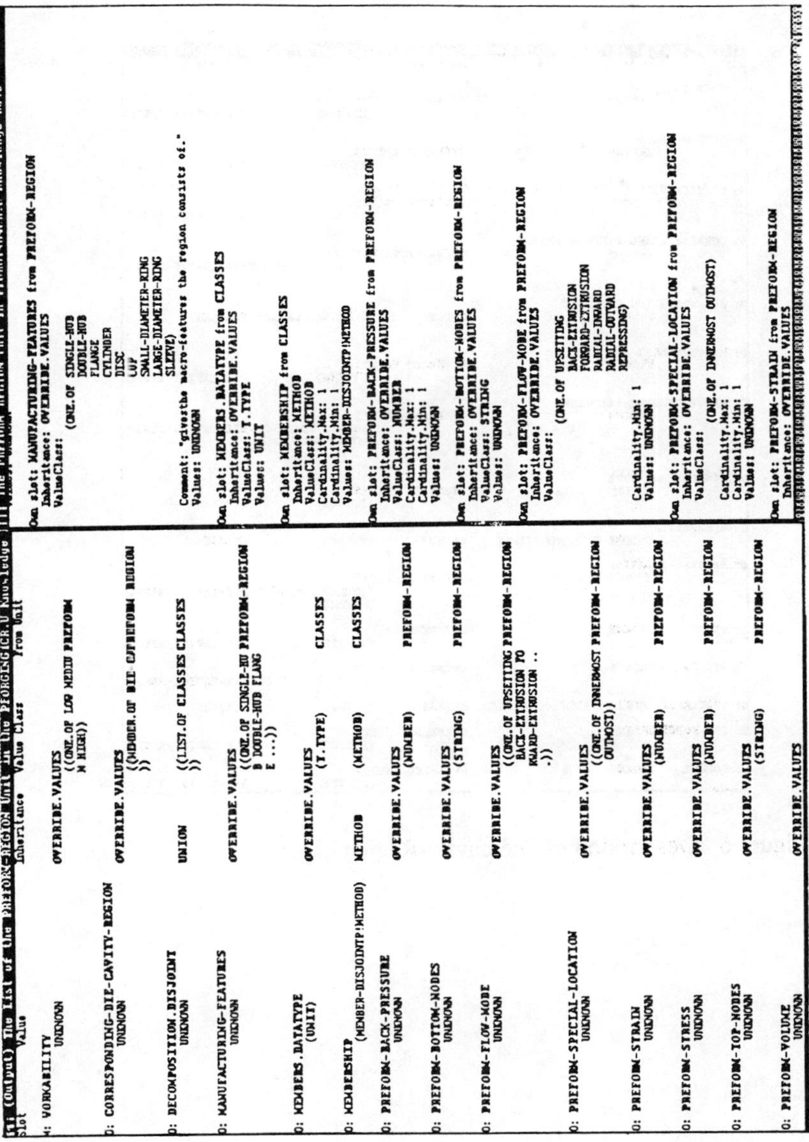

```
[1] (Output) The List of the PREFORM-REGION Unit in the PFORGINGICR.U Knowledge
Slot          Value                   Inheritance   Value Class      From Unit

M: WORKABILITY
              UNKNOWN                 OVERRIDE.VALUES
                                                    ((ONE.OF LOW MEDIU PREFORM
                                                     M HIGH))

O: CORRESPONDING-DIE-CAVITY-REGION
              UNKNOWN                 OVERRIDE.VALUES
                                                    ((MEMBER.OF DIE-C/PREFORM REGION
                                                     ))

O: DECOMPOSITION.DISJOINT
              UNKNOWN                 UNION         ((LIST.OF CLASSES CLASSES
                                                     ))

O: MANUFACTURING-FEATURES
              UNKNOWN                 OVERRIDE.VALUES
                                                    ((ONE.OF SINGLE-HU PREFORM-REGION
                                                     B DOUBLE-HUB FLANG
                                                     E ...))

O: MEMBERS.DATATYPE
              (UNIT)                  OVERRIDE.VALUES
                                                    (T.TYPE)          CLASSES
O: MEMBERSHIP
              (MEMBER-DISJOINTP!METHOD)  METHOD     (METHOD)          CLASSES

O: PREFORM-BACK-PRESSURE
              UNKNOWN                 OVERRIDE.VALUES
                                                    (NUMBER)          PREFORM-REGION

O: PREFORM-BOTTOM-NODES
              UNKNOWN                 OVERRIDE.VALUES
                                                    (STRING)          PREFORM-REGION

O: PREFORM-FLOW-MODE
              UNKNOWN                 OVERRIDE.VALUES
                                                    ((ONE.OF UPSETTING PREFORM-REGION
                                                     BACK-EXTRUSION FO
                                                     RWARD-EXTRUSION ..
                                                     .))

O: PREFORM-SPECIAL-LOCATION
              UNKNOWN                 OVERRIDE.VALUES
                                                    ((ONE.OF INNERMOST PREFORM-REGION
                                                     OUTMOST))

O: PREFORM-STRAIN
              UNKNOWN                 OVERRIDE.VALUES
                                                    (NUMBER)          PREFORM-REGION

O: PREFORM-STRESS
              UNKNOWN                 OVERRIDE.VALUES
                                                    (NUMBER)          PREFORM-REGION

O: PREFORM-TOP-NODES
              UNKNOWN                 OVERRIDE.VALUES
                                                    (STRING)          PREFORM-REGION

O: PREFORM-VOLUME
              UNKNOWN                 OVERRIDE.VALUES
```

```
[1] The PREFORM REGION Unit in PFORGING4CR.U Knowledge Base

Own slot: MANUFACTURING-FEATURES from PREFORM-REGION
    Inheritance: OVERRIDE.VALUES
    ValueClass:
                (ONE.OF SINGLE-HUB
                        DOUBLE-HUB
                        FLANGE
                        CYLINDER
                        DISC
                        CUP
                        SMALL-DIAMETER-RING
                        LARGE-DIAMETER-RING
                        SLEEVE)
    Comment: "givesthe macro-features the region consists of."
    Values: UNKNOWN

Own slot: MEMBERS.DATATYPE from CLASSES
    Inheritance: OVERRIDE.VALUES
    ValueClass: T.TYPE
    Values: UNIT

Own slot: MEMBERSHIP from CLASSES
    Inheritance: METHOD
    ValueClass: METHOD
    Cardinality.Max: 1
    Cardinality.Min: 1
    Values: MEMBER-DISJOINTP!METHOD

Own slot: PREFORM-BACK-PRESSURE from PREFORM-REGION
    Inheritance: OVERRIDE.VALUES
    ValueClass: NUMBER
    Cardinality.Max: 1
    Cardinality.Min: 1
    Values: UNKNOWN

Own slot: PREFORM-BOTTOM-NODES from PREFORM-REGION
    Inheritance: OVERRIDE.VALUES
    ValueClass: STRING
    Values: UNKNOWN

Own slot: PREFORM-FLOW-MODE from PREFORM-REGION
    Inheritance: OVERRIDE.VALUES
    ValueClass:
                (ONE.OF UPSETTING
                        BACK-EXTRUSION
                        FORWARD-EXTRUSION
                        RADIAL-INWARD
                        RADIAL-OUTWARD
                        REPRESSING)
    Cardinality.Min: 1
    Values: UNKNOWN

Own slot: PREFORM-SPECIAL-LOCATION from PREFORM-REGION
    Inheritance: OVERRIDE.VALUES
    ValueClass:
                (ONE.OF INNERMOST OUTMOST)
    Cardinality.Max: 1
    Cardinality.Min: 1
    Values: UNKNOWN

Own slot: PREFORM-STRAIN from PREFORM-REGION
    Inheritance: OVERRIDE.VALUES
```

Figure 5 - Description of Preform-Region object

```
### The List of the DIE-CAVITY-REGION Unit in the PIORGING4CH.U Knowledge Base
Slot            Value                         Inheritance     Value Class      From Unit

O: BACK-PRESSURE
                UNKNOWN                       OVERRIDE.VALUES
                                                              (NUMBER)         DIE-CAVITY-REGION

O: BOTTOM-NODES
                UNKNOWN                       OVERRIDE.VALUES
                                                              (STRING)         DIE-CAVITY-REGION

O: CONNECTIVITY
                UNKNOWN                       OVERRIDE.VALUES
                                                              (STRING)         DIE-CAVITY-REGION

O: CORRESPONDING-PREFORM-REGION
                UNKNOWN                       OVERRIDE.VALUES
                                                              ((MEMBER.OF PREFORM DIE-CAVITY-REGION
                                                              ))

O: DECOMPOSITION.DISJOINT
                UNKNOWN                       UNION           ((LIST.OF CLASSES) CLASSES
                                                              )

O: HAS-SURFACES
                UNKNOWN                       OVERRIDE.VALUES
                                                              (STRING)         DIE-CAVITY-REGION

O: MANUFACTURING-FEATURES
                UNKNOWN                       OVERRIDE.VALUES
                                                              ((ONE.OF SINGLE-HUB DIE-CAVITY-REGION
                                                              DOUBLE-HUB FLANGE
                                                              ...))

O: MEMBERS.DATATYPE
                (UNIT)                        OVERRIDE.VALUES
                                                              (T.TYPE)         CLASSES

O: MEMBERSHIP
                (MEMBER-DISJOINTP!METHOD)     METHOD          (METHOD)         CLASSES

O: SPECIAL-LOCATION
                UNKNOWN                       OVERRIDE.VALUES
                                                              ((ONE.OF INNERMOST DIE-CAVITY-REGION
                                                              OUTMOST))

O: STRAIN       UNKNOWN                       OVERRIDE.VALUES
                                                              (NUMBER)         DIE-CAVITY-REGION

O: STRESS       UNKNOWN                       OVERRIDE.VALUES
                                                              (NUMBER)         DIE-CAVITY-REGION

O: SUBCLASSP (SUBCLASS-DISJOINTP!METHOD)      METHOD          (METHOD)         CLASSES

O: TOP-NODES UNKNOWN                          OVERRIDE.VALUES
                                                              (STRING)         DIE-CAVITY-REGION

O: VOLUME       UNKNOWN                       OVERRIDE.VALUES
                                                              (NUMBER)         DIE-CAVITY-REGION
```

Figure 6 - Description of Die-cavity-Region object

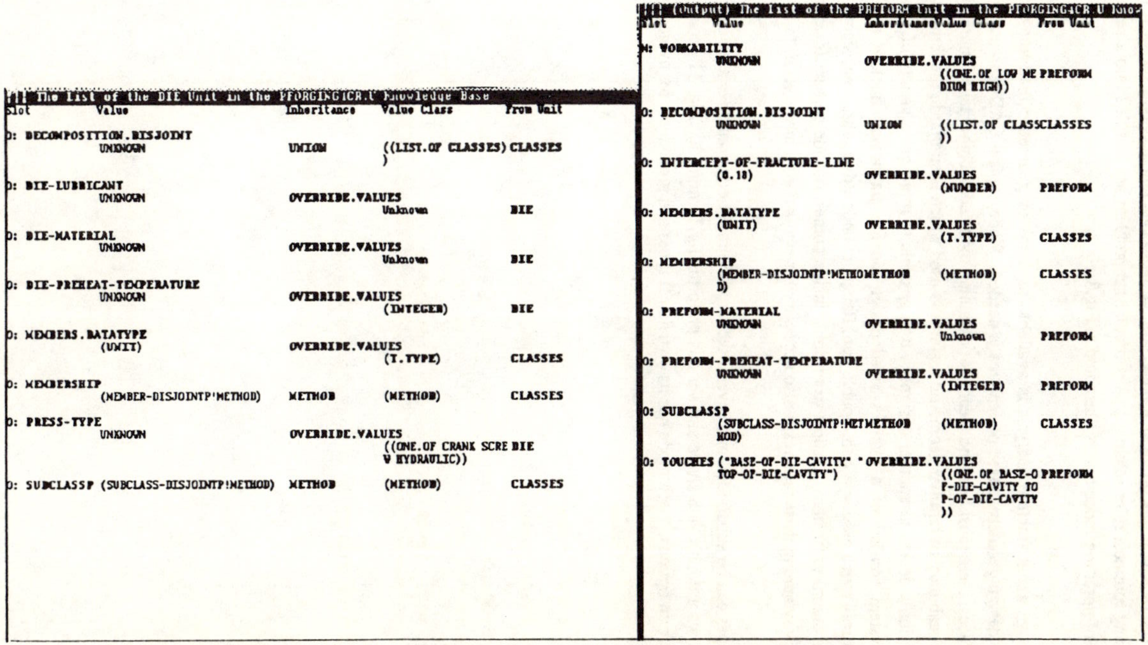

Figure 7 - Description of Die and Preform object

Implementation of the methodology provides access to mathematical models of plastic deformation. Procedural information is expressed in LISP language. There is a natural language-like interface for two-way interaction with the knowledge base. Reasoning is done with decision rules which explicitly describe the context in which they are applicable. Rules are a declarative way of expressing unordered knowledge. About fifty rules have been formulated for reasoning about Free surface fracture, Die-contact fracture, Internal defects and Corner lifting. Different decision rules may apply to each flow mode in combination with the location and geometry of the region. For the purpose of structuring problem solving, ease of debugging and to reduce the CPU time consumed during pattern matching routines, rules relating to a particular subproblem are bundled into a seperate class. An example of a section of upsetting-mode rule is illustrated in Fig.8.

Most of the knowledge about fracture analysis is available in terms of geometric features; thus, a symbolic representation of the design geometry based on features is needed. Since, typical CAD representations consist of nodes, lines and surfaces, there must be some means for obtaining the needed higher level (more abstract) feature representations. A vocabulary of familiar geometric features is used which is similar to the terms with which one designer would describe a design to another designer. It is possible to represent a variety of designs using this vocabulary. The system consists of a features-based design aid for diagnosing fracture sites. It employs **manufacturing features** (such as single hub, double hub, flange, cylinder, disc, cup, ring, sleeve) together with **co-features** (such as lateral extensions, vertical extensions, inside corner, outside corner) for both preform and die-cavity. Connectivity among these features is also necessary for evaluation of manufacturability.

Complex forged shapes can be subdivided into several sections of simplified geometries as shown in Fig. 10 and Fig. 11. Fig. 10 shows a double hub shape and various preforms for forming it. Fig. 11 shows back extruded shapes and some preform options. Each mode has its own defect types and procedures for avoiding defects.

```
(UPSETTINGMODE-RULE
 (IF (?D-R IS IN CLASS DIE-CAVITY-REGION)
      (THE MANUFACTURING-FEATURES OF ?D-R IS SINGLE-HUB)
      (THE MANUFACTURING-FEATURES OF ?D-R IS FLANGE)
      (THE CORRESPONDING-PREFORM-REGION OF ?D-R IS ?P-R)
      (THE FLOW-MODE OF ?P-R IS UPSETTING)
      (THE DIAMETER
           OF
           PREFORM-EXTENSIONS
           IS
           ?DIA-P)
      (THE DIAMETER OF DIE-EXTENSIONS IS ?DIA-D)
      (THE HEIGHT OF PREFORM-EXTENSIONS IS ?H-P)
      (THE HEIGHT OF DIE-EXTENSIONS IS ?H-D)
      (THE FRICTION OF DIE-CAVITY-SURFACES IS ?M)
      (THE H/D OF PREFORM-EXTENSIONS IS ?H/D)
      (THE INTERCEPT-OF-FRACTURE-LINE
           OF
           PREFORM
           IS
           ?A1)
      (LISP
        (LET* ((KA
                (+ -4.37
                   (* ?M 18.28)
                   (* (EXPT ?M 2) -28.4)
                   (* (EXPT ?M 3) 14.08)))
               (KB
                (+ 5.96
                   (* ?M -30.82)
                   (* (EXPT ?M 2) 47.46)
                   (* (EXPT ?M 3) -22.95)))
               (MA
                (+ 1.33
                   (* ?M -10.72)
                   (* (EXPT ?M 2) 16.59)
                   (* (EXPT ?M 3) -8.26)))
               (MB
                (+ -3.51
                   (* ?M 23.25)
                   (* (EXPT ?M 2) -38.55)
                   (* (EXPT ?M 3) 20.18)))
               (A (- (* -1 MA ?H/D) KA))
               (B (- (* MB ?H/D) KB))
               (MAX-DIA-EXPANSION . . .
```

Figure 8 - Description of an example rule in the knowledge base

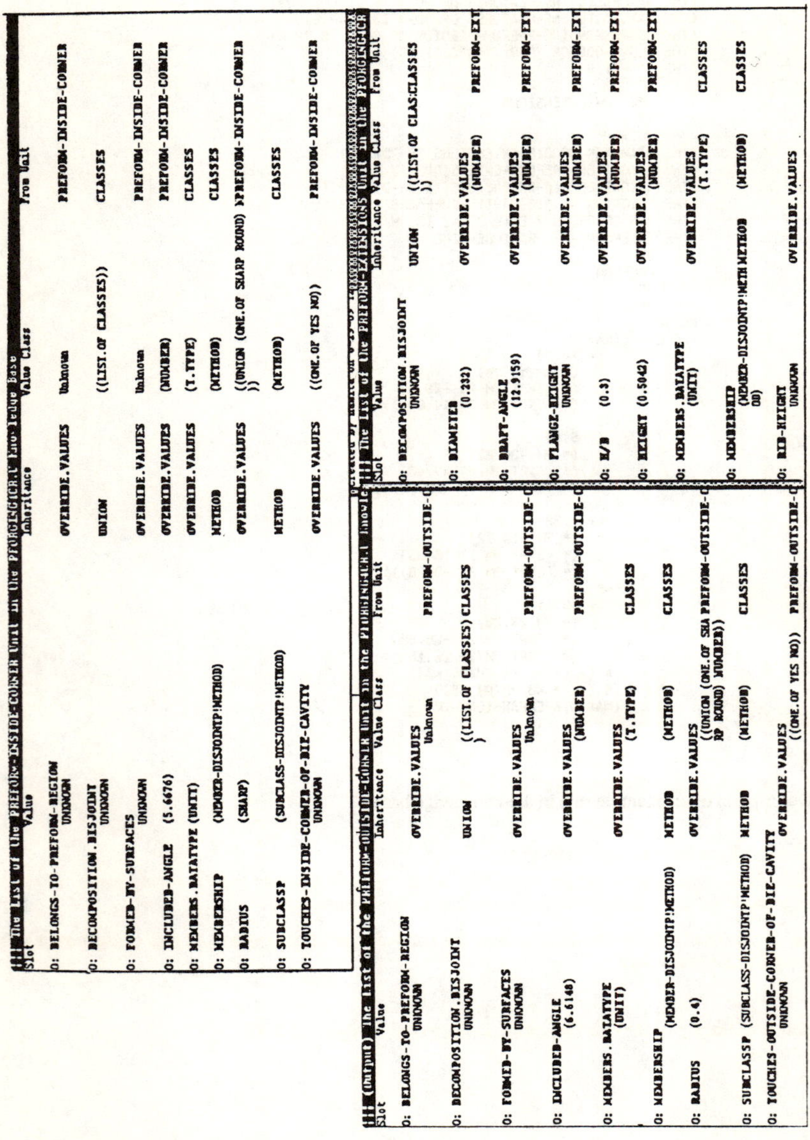

The List of the PREFORM-INSIDE-CORNER Unit in the PFORGING4CR.L Knowledge Base

Slot	Value	Inheritance	Value Class	From Unit
O: BELONGS-TO-PREFORM-REGION	UNKNOWN	OVERRIDE.VALUES	Unknown	PREFORM-INSIDE-CORNER
O: DECOMPOSITION.DISJOINT	UNKNOWN	UNION	((LIST.OF CLASSES))	CLASSES
O: FORMED-BY-SURFACES	UNKNOWN	OVERRIDE.VALUES	Unknown	PREFORM-INSIDE-CORNER
O: INCLUDED-ANGLE	(5.6676)	OVERRIDE.VALUES	(NUMBER)	PREFORM-INSIDE-CORNER
O: MEMBERS.DATATYPE	(UNIT)	OVERRIDE.VALUES	(T.TYPE)	CLASSES
O: MEMBERSHIP	(MEMBER-DISJOINTP!METHOD)	METHOD	(METHOD)	CLASSES
O: RADIUS	(SHARP)	OVERRIDE.VALUES	((UNION (ONE.OF SHARP ROUND)))	PREFORM-INSIDE-CORNER
O: SUBCLASSP	(SUBCLASS-DISJOINTP!METHOD)	METHOD	(METHOD)	CLASSES
O: TOUCHES-INSIDE-CORNER-OF-DIE-CAVITY	UNKNOWN	OVERRIDE.VALUES	((ONE.OF YES NO))	PREFORM-INSIDE-CORNER

(Output) The List of the PREFORM-OUTSIDE-CORNER Unit in the PFORGING4CR.L Knowle...

Slot	Value	Inheritance	Value Class	From Unit
O: BELONGS-TO-PREFORM-REGION	UNKNOWN	OVERRIDE.VALUES	Unknown	PREFORM-OUTSIDE-C
O: DECOMPOSITION.DISJOINT	UNKNOWN	UNION	((LIST.OF CLASSES))	CLASSES
O: FORMED-BY-SURFACES	UNKNOWN	OVERRIDE.VALUES	Unknown	PREFORM-OUTSIDE-C
O: INCLUDED-ANGLE	(6.6148)	OVERRIDE.VALUES	(NUMBER)	PREFORM-OUTSIDE-C
O: MEMBERS.DATATYPE	(UNIT)	OVERRIDE.VALUES	(T.TYPE)	CLASSES
O: MEMBERSHIP	(MEMBER-DISJOINTP!METHOD)	METHOD	(METHOD)	CLASSES
O: RADIUS	(0.4)	OVERRIDE.VALUES	((UNION (ONE.OF SHARP ROUND) NUMBER))	PREFORM-OUTSIDE-C
O: SUBCLASSP	(SUBCLASS-DISJOINTP!METHOD)	METHOD	(METHOD)	CLASSES
O: TOUCHES-OUTSIDE-CORNER-OF-DIE-CAVITY	UNKNOWN	OVERRIDE.VALUES	((ONE.OF YES NO))	PREFORM-OUTSIDE-C

The List of the PREFORM-EXTENSIONS Unit in the PFORGING4CR

Slot	Value	Inheritance	Value Class	From Unit
O: DECOMPOSITION.DISJOINT	UNKNOWN	UNION	((LIST.OF CLASSES))	CLASSES
O: DIAMETER	(0.232)	OVERRIDE.VALUES	(NUMBER)	PREFORM-EXT
O: DRAFT-ANGLE	(12.9159)	OVERRIDE.VALUES	(NUMBER)	PREFORM-EXT
O: FLANGE-HEIGHT	UNKNOWN	OVERRIDE.VALUES	(NUMBER)	PREFORM-EXT
O: H/B	(0.3)	OVERRIDE.VALUES	(NUMBER)	PREFORM-EXT
O: HEIGHT	(0.5042)	OVERRIDE.VALUES	(NUMBER)	PREFORM-EXT
O: MEMBERS.DATATYPE	(UNIT)	OVERRIDE.VALUES	(T.TYPE)	CLASSES
O: MEMBERSHIP	(MEMBER-DISJOINTP!METHOD)	METHOD	(METHOD)	CLASSES
O: RIB-HEIGHT	UNKNOWN	OVERRIDE.VALUES		

Figure 9 - Description of various Preform-Cofeatures.

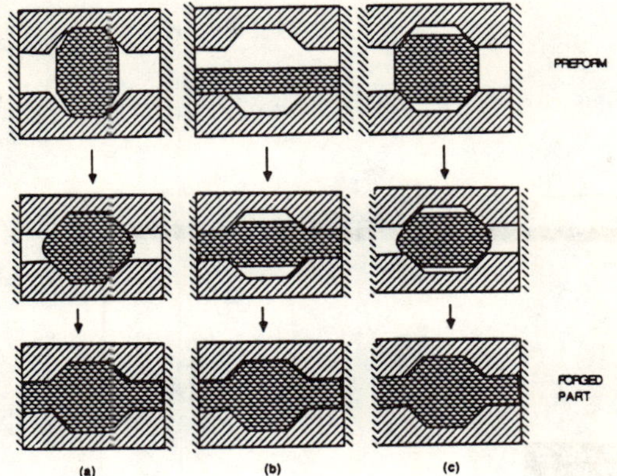

Figure 10 - Preform options for forging a flanged hub shape

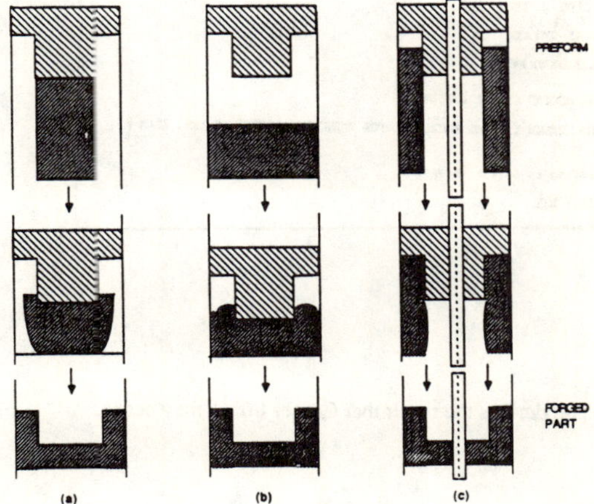

Figure 11 - Preform options for forging a cup or rim

Chaining over rules is invoked by two modes:

- Data driven reasoning to obtain results on what types of fracture would occur, given a set of processing conditions and geometric parameters of die and preform.

- Goal driven reasoning to obtain results on the combination of geometric and processing conditions required to avoid fracture.

An illustration of tracing of the chaining process is depicted in figure 12 & 13.

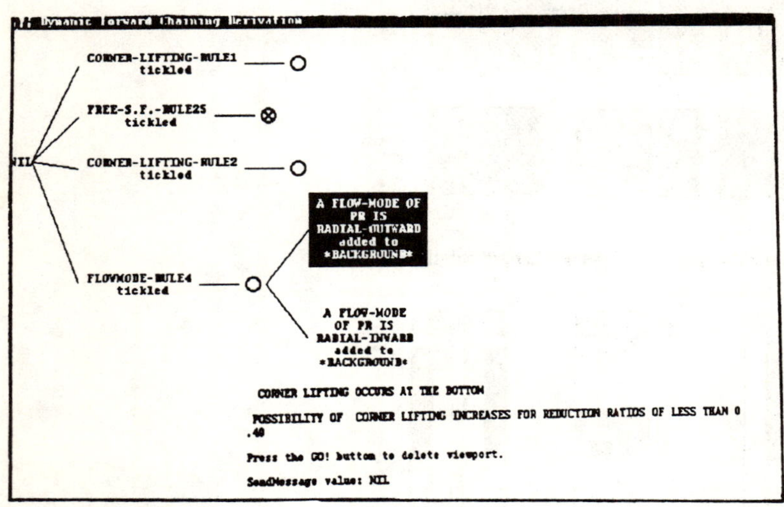

Figure 12 - Illustrates the reasoning process and derives the result that Corner lifting may occur.

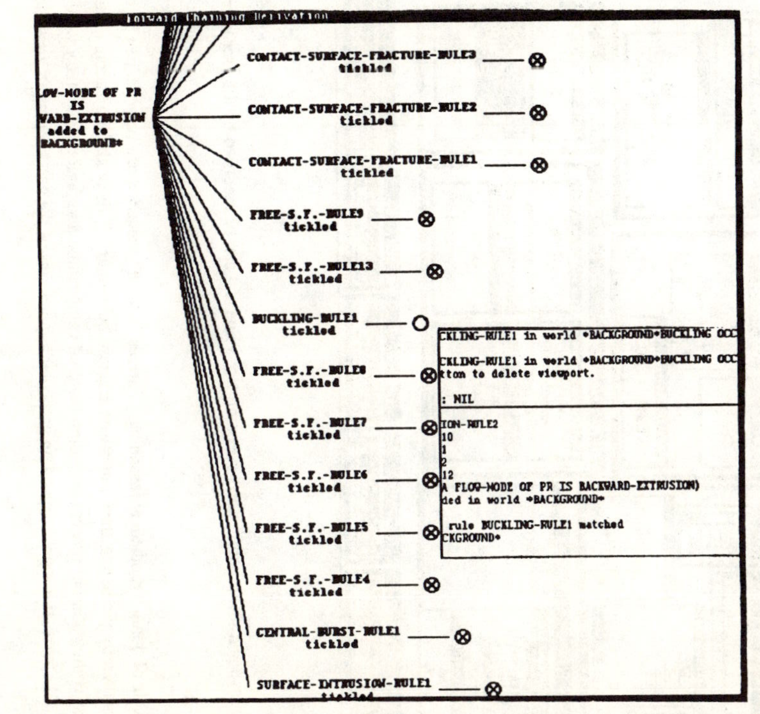

Figure 13 - Illustrates the reasoning process and arrives at the conclusion that Buckling may occur in the forged part. It also displays the various subgoals in the deriving process.

The decision support system can handle a wide variety of simple shapes covering all axisymmetric parts and axisymmetric parts with extensions, and characteristic flow modes such as backward extrusion, forward extrusion, radial inward, radial outward, upsetting and repressing.

The application interface is very user-friendly and enables easy access to the system, even for a novice user. At the same time, the implementation details of the system are transparent to the user. As shown in fig. 14, icons are used to input values of parameters for characterizing *extensions, inside-corner, outside-corner* of both die-cavity and preform. Menus are provided for *manufacturing features* of the die-cavity and preform which are capable of accepting multiple values by the user. Also, there is a selection menu for *flow-mode and "touches"* attribute of the preform. The value of the *intercept of the fracture line* can be input or changed by the user.

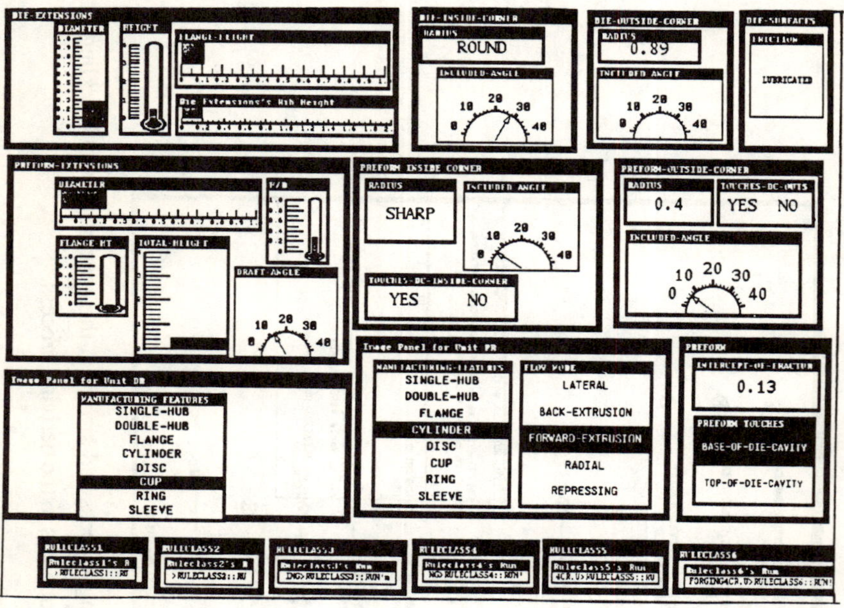

Figure 14 - Each guage and menu both reports the parameter value and can be modified by the user to reflect the updated value.

The system is capable of doing hypothetical reasoning. Various "what if" scenarios can be generated by changing the key parameters thereby predicting their impact on fracture behaviour. This leads to a rationalization as to whether a particular combination of process variables, material composition and preform geometry would lead to a defect-free part. Of course, this would depend on particular applications of the part keeping in view the desired mechanical properties.

Conclusions and Recommendations

Using the proposed architecture and diagnosis strategy desired, the fracture determining process is automated. This problem was initially ill-structured and complicated, but using a hybrid collection of AI development tools, an effective specification and solution is accomplished for relatively simple two-dimensional shapes.

In spite of all this, it is not yet clear whether a rich enough set of features can be provided to enable design of all complex parts without an explosion in the number of primitives. Flexibility exists in the system for a designer to add, delete and modify the existing knowledge base as needed for their design needs. If large amounts of information must be accessed, using a database management system is a good choice. This can be implemented by building mapping between commercial databases and knowledge bases to enable efficient bidirectional movement of information.

A concept of growing interest in manufacturing is *Concurrent Engineering*, which is a design methodology wherein a number of distinct aspects of a product are engineered simultaneously leading to elimination of inconsistencies and associated cost-factors. For instance, engineering different phases in a product life-cycle, such as conceptual design, detailed design, manufacturing process, planning, tooling, maintenance, cost etc. is approached at the same time in an integrated manner (14). Thus, efforts in design and manufacturing go hand in hand to avoid conflict and inconsistencies.

Further research is needed both to understand and develop more powerful and flexible knowledge based systems to be integrated with CAD systems and FEM techniques.

Acknowledgements

The authors extend their appreciation to the National Center of Excellence in Metalworking Technology, Johnstown, PA, for their support of this work.

References

1. H.A. Kuhn, and G.E. Dieter, <u>Advances in Research on the strength and fracture of Materials</u> (Pergamon Press, 1977), 307-324.

2. S.K. Suh, "Prevention of defects in Powder Preform Forging" (Ph. D. thesis, Drexel University, 1976).

3. B.L. Ferguson, H.A. Kuhn, and J. Trasorras, "Expert Software Concepts for Design of P/M Preforms for Precision Forging", <u>J. of Materials Shaping Technology, ASM</u>, 5 (1987), 97-105.

4. H.A. Kuhn, "Workability Theory and Application,"<u>Metals Handbook</u>,14 (Metals Park, OH:American Society for Metals, 1987), 388.

5. J. Shah, and H.A.Kuhn, "An empirical Formula for Workability limits in Cold Upsetting and Bolt Heading," Proc. North American Manufacturing Research Inst. Conference, SME, Berkeley, (1985) .

6. Toyohiko Okamoto, Takashi Fukuada, and Hyoji Hagita, "Material Fracture in Cold Forging - Systematic Classification of Working Methods and Types of Cracking in Cold Forging," <u>The Sumitomo Search</u>, 9 (1973), 216-226.

7. Y.C.Lee, and K.S.Fu, "Machine Understanding of CSG: Extraction and Unification of Manufacturing Features," <u>IEEE CG&A</u>, 1(1987), 20-32.

8. Intellicorp, Intellicorp KEE Software Development System Users Manual, <u>intellicorp</u>.

9. Geoffrey Goldbogen et. al., "Expert System for Extracting Manufacturing Features from a CAD database," <u>Expert Systems and Intelligent Manufacturing</u> (New York, NY:Elsevier Science Publishing Co., Inc., 1988), 285-304.

10. P.K. Wright, and D.A.Bourne,<u>Manufacturing Intelligence</u>, (New York, NY: Addison-Wesley Publishing Company, Inc., 1988).

11. Robert Hawley, <u>Artificial Intelligence Programming Environments</u> (New York, NY: John Wiley & Sons, 1987).

12. S.C.Y. Lu, and R. Kumanduri, <u>Knowledge-Based Expert Systems for Manufacturing</u> (Presented at the Winter Annual Meeting of the ASME, Anaheim, California, 1986). 311

13. P.W.Lee, and H.A. Kuhn, "Fracture in Cold Upset Forging - A criterion and model," <u>Met. Trans. A</u>, 4 (1973), 33-39.

14. Alexander Kott, "Concurrent Engineering" (Report by Carnegie Group Inc.,1989).

15. M.G. Cockroft, "Ductile Fracture in Cold working Operations," <u>Ductility</u>, (Metals Park, OH: American Society for Metals, 1967), 199-225.

DEVELOPMENT OF A FRACTURE ANALYSIS EXPERT SYSTEM USING BASIC CONCEPTS OF KNOWLEDGE ENGINEERING

Ma Cheng and Glen A. Stone

South Dakota School of Mines and Technology
501 East Saint Joseph Street
Rapid City, South Dakota 57701

ABSTRACT

This paper describes what is expected of an expert system, introduces the basic concepts and skills of knowledge engineering, and discusses the role of knowledge engineering and domain experts in developing expert systems. In particular, the organization of the knowledge base, the representation of the inference algorithm and organization of the rules are presented. The paper emphasizes the role of domain experts in developing expert systems and focuses on the principles underlying the design of expert systems. The method used to develop a user friendly interface is described. An example "Fracture Analysis Expert System" is introduced to illustrate these principles.

Expert System Applications in
Materials Processing and Manufacturing
Edited by M.Y. Demeri
The Minerals, Metals & Materials Society, 1989

INTRODUCTION
Expert System and Knowledge Engineering

Chris Shipley[1] defines *Expert System* as a computer program capable of considering a vast body of knowledge, reasoning, and then recommending a course of action.   Expert systems are different from conventional computer applications in that they are designed to simulate domain expert performance.   Therefore, expert systems are capable of making informed judgements using rules of thumb adopted from domain experts, and reaching (making) meaningful conclusions (decisions) when faced with complex or ambiguous information.[2]   Throughout almost two decades of experience, research showed that to solve problems in the area of human expertise such as engineering, medicine, or programming, the expert system programmers need to know what domain experts know about the subject; to make a fast and consistent symbol processor perform as well as a human expert, the domain expert must provide the computer specialized know-how comparable to what a human expert possesses.[3]   On the other hand, conventional software technique can not meet the requirements to imitate the experts problem-solving ability; to encode a significant amount of knowledge; to include facts, concepts, rules, strategies and the hierarchical relationships these elements imply.[2]

These needs give rise to *knowledge engineering*.   The technology of knowledge engineering is to provide the variety of techniques to capture, store, distribute, apply knowledge accurately as well as electronically, and to represent facts,   rules and   strategies accurately as well as consistently.[3] It has become apparent that "expert system development requires knowledge engineering"[2]. However, it is not necessary that expert system developers have to be knowledge engineers.   Actually with the aid of an expert system "shell" and some training, domain experts can not only transfer their knowledge to   knowledge engineers, but also expert system developers.   For example, a failure analysis expert system model presented by Volker Weiss[4] was "conceived by an expert in failure analysis, not a knowledge engineer."

Domain experts are different from knowledge engineers in that they know exactly the ways to solve task-specific problems.   Their process of developing an expert system is simpler than that of the knowledge engineer.   However, the problem is that most of the domain experts lack the background of knowledge engineers.   Therefore, how to form a knowledge model and build a prototype becomes an obstacle for the domain expert.

Some concepts and techniques used in developing our fracture analysis expert system are basic in developing any expert system.

Therefore, the discussion that follows is focused on introducing and discussing knowledge engineering principals in an effort to provide the basic idea to all domain experts who are interested in this area.

FRACTURE ANALYSIS EXPERT SYSTEM (FAES)
Basic Structure of Expert Systems

Generally, an expert system consists of a knowledge base, an inference engine and an user interface(Fig. 1). The *knowledge base* is an organized set of data structures which represent and store what the domain expert believes. The *user interface* provides different ways to communicate between program and end-user. Between the knowledge base and the user interface there is an *inference engine,* a program responsible for accepting beliefs, interpreting beliefs, and performing certain degrees of inference.[5]

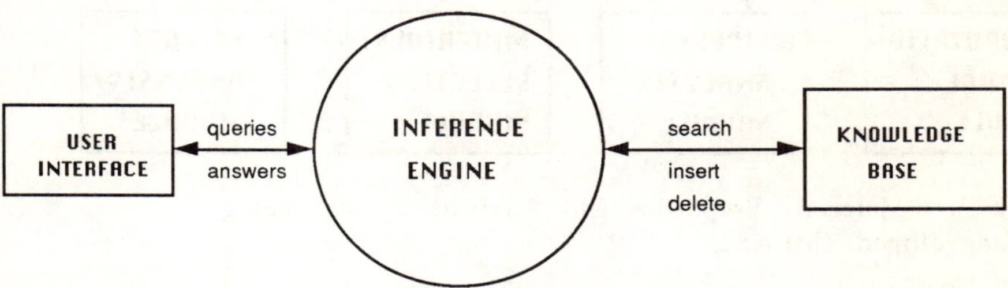

Figure 1.   Schematic of Expert System

FAES presented here is actually a module of "Materials Properties and Analysis System" (MPAS) which is currently being developed by the authors (Fig. 2) and will be described in more detail later. By providing macroscopic fracture information, obtained through visual and low magnification inspection of the surface, FAES is able to reason to its conclusion, a possible fracture mechanism and answers users questions.

FAES Knowledge Base
Facts and Rules

To design a knowledge base for any kind of task, the programmer starts by collecting and extracting the knowledge in solving the specific task. The knowledge is then encoded in an appropriate manner that presents the knowledge explicitly.[6] The knowledge base in FAES consists of general information about fracture analysis in the form of facts and rules.

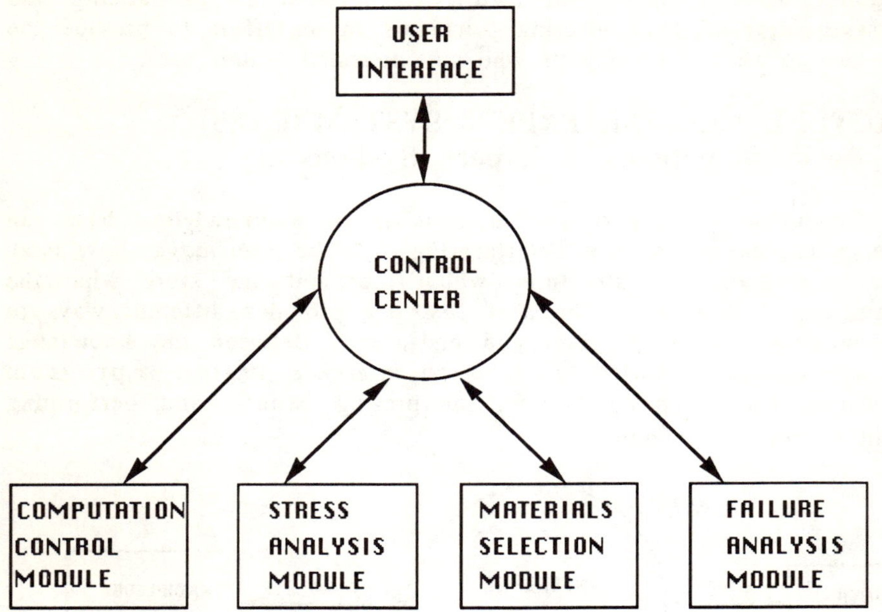

Figure 2.   Materials Properties and Analysis System being developed (MPAS).

A *fact*   refers to the set of possible states of affairs.[5] Facts allow the expression of arbitrary relationships between objects.   Suppose the fact expressed in prose **"the fracture has gross plastic deformation"** needs to be put into the knowledge base.   In PROLOG, it must be written in the form:

```
has(fracture,gross_plastic_deformation).
```

A key idea behind this form is that every expression denotes something **that is**.   There is some entity, in the example **fracture**, that each fact expression is associated.   Once the denotation of a fact is determined, as in the above example, it is fixed.[5]

A *rule*   is a fact which depends upon a group of other facts.[7] The program must know the relationship between the group of facts. Therefore a rule must be devised. In our implementation we utilize PROLOG.[7]   Consider the following example of a rule together with possible English interpretation.   The English expression **"the fracture mechanism is brittle if the fracture does not have gross plastic deformation"** is a potential rule. For simplicity this example of a rule involves only one fact, the fact illustrated above.   The syntax of the rule in PROLOG is:

```
is(fracture_mechnism,brittle)   :-
    not(has(fracture,   gross_plastic_deformation)).
```

If the clause of not(has(fracture,gross_plastic_deformation)) is
TRUE then the expert system may conclude that the fracture mechanism
is **brittle** in the absence of any other facts.

The knowledge base is controlled by the requirements of the
inference engine, and is used to encode and store information which may
expand and contract during program execution, generally, it should have
the following properties. 1. Easy for inference engine to search facts and
rules. 2. Easy to make incremental or decremental improvements. Those
features make expert system programs really different from
conventional programs.

PROLOG has facilities for searching and updating of the
knowledge base.   By placing often used information at the top of a
knowledge base, and inserting or deleting information dynamically
during execution, FAES can search its space effectively and manage
computer memory economically.

Certainty Factors

When developing a diagnostic expert system like the FAES,
the construction of the performance rules involves a difficult compromise
between the efficiency and the accuracy.   There is no universal approach
to solve real life problems.   Even for a specific task, different domain
experts can use different reasoning strategies to solve problems.

Therefore, some expert system developers prefer to use the
concept of *certainty factors* to estimate the accuracy of performance rules
and conclusions.   Since the focus of the paper is mainly on the
introduction of the basic concepts of expert systems and knowledge
engineering, the concept of certainty factors is not included in FAES.
Although, this does not imply that the concept is unimportant.

FAES Inference Engine

Recall that an *inference engine* is a program responsible for
accepting beliefs, interpreting beliefs, and performing certain degrees of
inference.   The inference engine algorithm is faced with the problem of
searching for solutions.   The inference engine should generate and test all
possibilities until a solution is found.   This set of possibilities is called the
search space.   This search algorithm used for that kind of application is
called *brute force*.[3] This kind of algorithm can be used only when the
search space is small.   As the search space expands (as it is in our

program MPAS) the efficiency of the searching must be improved. The efficiency can be improved by eliminating from consideration collective solutions, that can not succeed due to the unique set of facts presented to the expert system. This type of search reduction is called *pruning*[3].

In Fig.3 a search space with eight elements of equal possibility is illustrated. Three "yes-no questions" are needed to remove all the uncertainty of the search space. The task is how to develop an inference algorithm whose job is to prune the search space. The structure that accomplishes this task can be represented as a binary tree. Fig. 3 is an example of the *binary tree* used by FAES to discover natures of fractures. Since three bits of information can remove all uncertainty. If the tree has more than three levels, then the algorithm will be considered less efficient, on the other hand, it is incomplete if the tree has less than three levels.

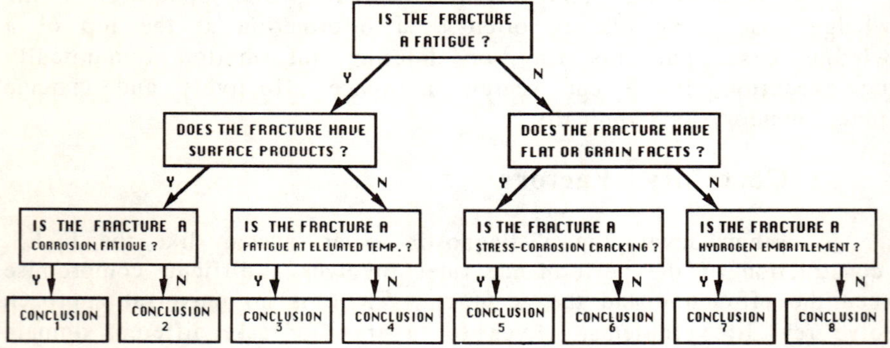

Figure 3.   Binary Tree for Discovering the Natures of a Fracture.

The creation of the tree in Fig. 3 could be based on the information associated with each node. The nodes on the left sub-tree correspond to positive answers. Similarly, the nodes on the right sub-tree correspond to negative answers. In either case the associated features are logically closer to their conclusion than the features associated with the root node of the tree (or sub-tree). At each level the number of possible conclusions is reduces by a factor of two. The search space is effectively reduced.

To search the space systematically, there are two basic methods: depth-first and breadth-first search. In a *depth-first* search, the inference engine produces a sub-goal at every opportunity, delving deeper and deeper into the details or sequences of rules. A *breadth-first* search, on the other hand, surveys all the premises for a rule before examining the details of the rule.[8] For a tree like Fig. 4, three "yes-no questions" are enough for depth-first search to remove all uncertainty of the problem; but for breadth-first search all seven "yes-no questions" are required to get the same amount of information. Therefore, the binary tree and depth-first search is the optimal method to search a space.

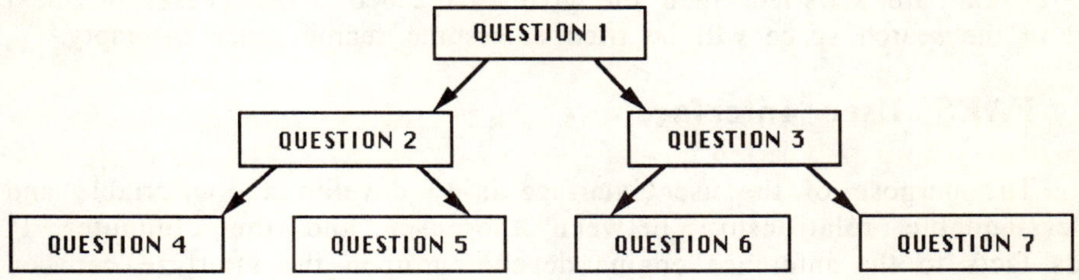

Figure 4.   Question Tree Combining Sets of Information.

Besides the depth-first method which is discussed above to search a space efficiently, a strategy is also required to provide a reasoning mechanism to search for a goal or a conclusion.   There are two commonly used reasoning mechanisms; they are forward and backward chaining. For a *forward chaining* mechanism, the inference starts from the facts which are randomly volunteered by the user.   The expert system's objective is to use this information to reach goals (or sub-goals).   On the other hand, for a *backward chaining* strategy, the search routine starts from possible goals, then it communicates with end-user to achieve appropriate facts.   The prerequisite for the forward chaining is that the volunteered facts must be presented in a specific manner that can be accepted by the program.   Many expert systems use combined backward chaining and forward chaining.[5 and 8]

For a simple implementation in PROLOG we have used FAES inference rules which exhibit a depth-first, backward-chaining control strategy.   By searching for sufficient confirming evidence, its reasoning line starts from potential fracture mechanisms in a search space.   The nearest elements in the search space is picked up as the target mechanism.   Then it queries facts according to the most likely and characteristic syndromes.   The target mechanism is tested until a disconfirming fact is encountered. The expert system then moves to the next nearest mechanism in the search space.   For example, to determine whether or not a fracture mechanism is due to thermal fatigue, the depth-first, backward-chaining inference rule may look like

```
is(fracture_mechanism,thermal_fatigue)   :-
    is(fracture,fatigue),
    are(there,corrosion_products_on_the_surface),
    is(amount_of_corrosion,inversely_proportional
            _to_the_depth_of_the_crack).
```

The strategy starts from the goal of "**the fracture mechanism of the sample is thermal fatigue.**" If all of the sub-goals "**fracture is fatigue**"; "**there are corrosion products on the surface**"; and "**amount of corrosion is inversely proportional to the depth of the crack**" are satisfied, then the goal is reached. Otherwise, the next goal in the search space will be tried unless the search space is empty.

FAES User Interface

The purpose of the user interface is to develop a comfortable and understandable relationship between the user and the computer. It feeds facts to the inference engine depending upon the interface between the inference engine and the end-user. It can also explain how the program has reached a conclusion, which is dependent upon rules and facts resident in the knowledge base. In FAES, the user interface is driven by the inference engine. When invoked by the inference engine, the user interface generates questions and dialogs with the end-user. Generally end-users will not be domain experts, therefore this method is reasonable and convenient.

Suppose the system has to know the fact "**whether or not the fracture has gross plastic deformation**", the code can be

```
ask(fracture,has,gross_plastic_deformation,Answer)   :-
    write("Does the fracture have gross plastic
    deformation  ?"),
    readln(Answer).
```

write is a built-in predicate in PROLOG for output, and **readln** is a built-in predicate to read input characters. The answer volunteered by the end-user is echoed to the parameter **Answer** which is in the predicate **ask**.

Because domain experts will be interested in both the conclusion and the reasoning line being used to reach the conclusion, it is necessary that the user interface communicates with the domain expert through dialog.   Therefore, besides the ability to inquire with "yes-no questions", FAES must be able to explain how it has reached the conclusion through the user interface unit.   The method being used here is to link the conclusion with the corresponding rules and facts.   In PROLOG, the predicate **append** can be used to link two lists together, therefore it is often used to support the explanation feature. The definition of **append** is:

```
append([  ],L,L).
append([X|L1],L2,[X|L3])  :-  append(L1,L2,L3).
```

For example, the fact "**the fracture has no gross plastic deformation**" is required to be linked with the conclusion "**the fracture is brittle**", the code can be

append(["the fracture is brittle"],["the fracture has no gross plastic deformation"],Conclusion_list).

There are many strategies which can be used to achieve different levels of interaction with the user interface. The method presented here illustrates how the user interface in FAES was constructed.

DISCUSSION

The current version of the FAES is presented in Appendix A. The program consults with the end-user by asking appropriate questions. The program then attempts to reach a conclusion based on the end-users' answers and the performance rules contained in the knowledge base. If a conclusion is reached, it is displayed on the screen. In addition the FAES can explain, if requested, how the conclusion was reached.

As discussed above, the FAES uses depth-first, backward chaining strategy. Because this type of systems needs the program queries with end-user to get facts, the success of the application is related to the quality of the user interface. The method being used in the FAES interface is to asks "yes-no questions" and wait for answers. It is understood that the knowledge of the end-user can vary widely. Frequently the end-user can be confused by the vocabulary used in the program. Therefore, the ability to explain different levels of technical terms is considered critical in a commercial expert system. For simplicity, the version of FAES presented in Appendix A does not contain this feature. To add this feature, a method similar to that used to explain how conclusion are reached, which is contained in the program, can be constructed.

FAES can help determine fracture mechanism. The possibilities are normal fatigue, corrosion fatigue, thermal fatigue, fatigue at elevated temperatures, stress-corrosion cracking, hydrogen embrittlement, liquid-metal embrittlement, and stress-rupture failure. An example of how an unskilled user could use an expert system such as FAES is illustrated though the use of a fracture shown in Fig. 5. A typical dialog with the expert system to analyze this fracture follows. Note that all requested facts require only visual data. All information generated by the expert system is presented in bold type. User responses are in normal type. The program starts by suggesting a goal.

Figure 5.   The fracture shown came from a limestone crusher drive shaft (6 inches in diameter).

FAES Dialog

GOAL: To determine if the fracture is corrosion fatigue.
Does the fracture have beach marks?   yes
Does the fracture have a surface corrosion product?   no
CONCLUSION: The fracture is not corrosion fatigue.

GOAL: To determine if the fracture is fatigue at elevated temperatures.
Does the fracture have a multiplicity of creep voids near it?   no
CONCLUSION: The fracture is not fatigue at elevated temperatures.

GOAL: To determine if the fracture is normal fatigue.
CONCLUSION: The fracture is normal fatigue.

Are you interested in how the conclusion was reached?   yes

      START OF EXPLANATION:
RULE 4: The fracture is normal fatigue if it has beach marks.
FACT: The fracture has beach marks.
      END OF EXPLANATION.

As discussed above, the end-user may be confused by the words such as fatigue, beach marks or other technical terms. Thus, the ability of explaining the technical terms is very important to commercial expert systems. A dictionary of technical terms is being developed for the MPAS.

Potential applications of diagnostic expert systems are tremendous. An expert system application can obtain high yields only if applied to very specific, highly valued, knowledge-intensive applications.[1] Therefore during the design phase, enough attention must be paid to prevent the expert system from becoming too complex. For example, CORDIAL[9] (CORrosion DIagnosis Intelligent system for ALuminum) developed in Alcoa Laboratories incorporates fifty years "scientific, heuristic, and experimental knowledge of corrosion experts and utilizes this knowledge to aid a user in the diagnosis of stress corrosion cracking behavior." In materials science and metallurgy, failure analysis, materials selection, materials design, and heat treatment techniques are just a few candidates for expert system applications.

The MPAS illustrated in Fig. 2 is currently being developed by the authors. The purpose of the expert system is to develop a tool for materials engineers as well as metallurgical engineers. It consists of four modules. The computation control module calculates all the stresses in different sections the part. The stress analysis module analyses these stresses and specifies the materials properties required for the part. The materials selection module provides recommendations of materials and fabrication methods which can be successfully used for the application. The failure analysis module will have two tasks: the one illustrated in this paper, and the task of evaluating the engineering design of the part and predicting failure modes. The FAES introduced here is actually part of one of the modules in the MPAS.

CONCLUSIONS

1. The FAES uses depth-first, backward chaining strategy. It communicated with end-users through "yes-no questions". It is able to explain how the program has reached its conclusion. Its knowledge base has dynamic features.

2. The concept of certainty factors has not been used in the FAES. During the design phase of an expert system, the necessity of applying the concept should be considered.

3. The definition of technical terms has been left of the version of FAES presented here in an effort the keep the program simple. However, in a commercial version a dictionary must be added.

4. Binary tree is presented here to represent the inference algorithm. Following the algorithm depicted by the binary tree, the interpretation of the logical relation to PROLOG code is relatively straight forward. By comparing the levels of the binary tree with the

result calculated from information theory[10], the efficiency of the inference algorithm can be easily estimated.

5. The authors believe that it is the time to attract the attention of domain experts and emphasize the role of domain experts in the development of expert systems.

REFERENCES

1. Chris Shipley, "Whatever Happened to AI ?", *PC Computing*, Vol. 2, No. 3, 1989, pp 64-74.
2. Paul Haley, Chuck Williams, "Expert System Development Requires Knowledge Engineering," *Computer Design,* Vol. 25, No. 4, 1986, pp 83-88.
3. Frederick Hayes-Roth, "The Knowledge-Based Expert System : A Tutorial," *Computer,* Vol. 17, No. 9, 1984, pp 11-28.
4. Volker Weiss, "A Material Failure Analysis Expert System," Proceedings of Symposium on Applications of Artificial Intelligence in Materials Science, *Artificial Intelligence Applications in Materials Science*, 1986, pp147-156.
5. Eugene Charniak, Drew McDermott, *Introduction to Artificial Intelligence,* Addison-Wesley Publishing Co., Massachusetts, 1987.
6. Tom M. Mitchell, Louis I. Steinberg, and Jeffrey S. Shulman, "A Knowledge-based Approach to Design," *IEEE Transactions,* Vol. PAMI-7,No.5,1985,pp502-510.
7. W.F. Clocksin and C.S. Mellish, *Programming in Prolog*, Springer-Verlag, Berlin Heidelberg, Germany, 2nd edition, 1984.
8. Ben Finkel, "Tapping into the Knowledge Power of Expert System," *Computer Design,* Vol. 25, No. 6, 1986, pp 95-98.
9. William A. Boag Jr., et al., "CORDIAL - A Knowledge-Based System for the Diagnosis of Stress Corrosion Behavior in High Strength Aluminum Alloys,"Proceedings of Symposium on Applications of Artificial Intelligence in Materials Science, *Artificial Intelligence Applications in Materials Science*, 1986, pp 123-146.
10. Jean-Paul Tremblay, Paul G. Sorenson, *An Introduction to Data Structures with Applications,* McGraw-Hill, Inc., 1984.

APPENDIX A

The following PROLOG program provides an example of one method to structure an expert system for a materials engineering application. The objective is to provide a framework from which other applications could be built. Throughout the program listing an effort to present the task of each section of the program has been provided. The following PROLOG code is TURBO PROLOG available from Borland International, 4585 Scotts Valley, CA 95066.

```prolog
/*                      FRACTURE.PRO                            */
/*              Fracture Analysis Expert System                 */
/*                       Spring, 1989                           */
/*                                                              */
/*                      DECLARATION                             */
```

/*The **DECLARATION** section of a turbo prolog program is responsible for defining predicates, arguments, goa. and predicates being used in database.*/

```prolog
domains
        sml=symbol      smllist=symbol*      str=string      strlist=str*
database
        has(str,str,strlist)      is_a(sml,str,strlist)      pass(str)
predicates
        aim(str)          append(strlist,strlist,strlist)
        ask(str,str,str,strlist)      ask_how(sml,str)
        blank      clrdbase      explain_how(str,sml,str)
        fracture      how(sml,str)      is(sml,str,strlist)
        member(str,strlist)      output(sml,str)
        question(str,str,strlist)      respond(str,str,str,strlist)
        rule1(strlist)      rule11(strlist)      rule12(strlist)      rule13(strlist)
        rule2(strlist)      rule21(strlist)      rule22(strlist)      rule23(strlist)
        rule3(strlist)      rule31(strlist)      rule32(strlist)
        rule4(strlist)      rule41(strlist)
        rule5(strlist)      rule51(strlist)      rule52(strlist)
        rule6(strlist)      rule61(strlist)      rule62(strlist)
        rule7(strlist)      rule71(strlist)
        rule8(strlist)      rule81(strlist)      rule82(strlist)
        search(strlist)      space(strlist)      sub_how(strlist)
        word(str,str,str)
goal
        fracture
```

```prolog
/*                        MAIN  UNIT                            */
```

/*The **MAIN UNIT** contains the search space, control rules and search rules.*/

```prolog
clauses
```

/* The purpose of the predicate **fracture** is to find the fracture's features.*/

```prolog
        fracture:-
                        clearwindow,  cursor(3,10),
                write("FAES can help determine fracture mechanism.  The"),
                cursor(4,8),
                write("possibilities are normal fatigue, corrosion fatigue,"),
                cursor(5,8),
                write("thermal fatigue, fatigue at elevated temperatures,"),
                cursor(6,8),
                write("stress-corrosion  cracking,  hydrogen  embrittlement,"),
                cursor(7,8),
                write("liquid-metal  embrittlement,  stress-rupture  failure."),
                nl, nl,  space(X), !,  search(X), !,
                        clrdbase, nl, blank,
                write("The program has been terminated.  Press ENTER to"),
                write("return to menu.\n").
```

/*The list in the **space** predicate contains all the possible fracture mechanisms being coded in the program.*/

```prolog
space(["corrosion    fatigue","thermal    fatigue",
    "fatigue   at   elevated   temperatures","normal   fatigue",
    "stress-corrosion   cracking","liquid-metal   embrittlement",
    "hydrogen   embrittlement","stress-rupture   failure"]).
```

/* The predicate **search** allows searching of the **space** list recursively.*/

```prolog
search([ ]):-
    blank,
    write("This program can not determine the fracture feature."),
    !, nl.
search([H|_]):-
    is(fracture,H,_).
search([H|T]):-
    pass("yes"), !, blank,
    retract(pass("yes")), asserta(pass("no")),
    write("CONCLUSION: The fracture is not a ",H,".\n"), nl,
    search(T).
search([_|T]):-
    search(T).
```

```
/ *                    PERFORMANCE UNIT                          * /
```

/*The **PERFORMANCE UNIT** contains the rules with the reasoning strategies and dynamic knowledge base techniques.*/

/* The predicate is(Fracture,Mechanism, Conclusion_list) determines if the **Fracture** matches some specific **Mechanism**. If it matches, then the conclusion, which is a list of appropriate rules and facts will be linked together and stored in the **Conclusion_list**.*/

```prolog
is(Fracture,"corrosion    fatigue",Conclusion_list):-
    aim("corrosion    fatigue"),
    ask("fracture"," has ","beach    marks",_),
    asserta(is_a(Fracture,"nomal    fatigue",["FACT:    The fracture
            has    beach    marks."])),
    ask("fracture"," has ","a surface    corrosion    product",D2),
    ask("fracture"," has ","multiplicity   of   crack   origins",D1),
    ask("corrosion    product"," is ","filling   the   multiplicity   of
            cracks",D3), !,
    rule1(Z0), rule11(Z1), rule12(Z2), rule13(Z3),
    append(Z0,Z1,T0), append(Z2,Z3,T1), append(T0,T1,Z),
    append(Z,D,L), append(L,D1,L1),
    append(L1,D2,L2), append(L2,D3,Conclusion_list),
    asserta(is_a(Fracture,"corrosion    fatigue",Conclusion_list)),
    output(Fracture,"corrosion    fatigue").

is(Fracture,"thermal    fatigue",Conclusion_list):-
    is_a(Fracture,"normal    fatigue",D),
    has("fracture","a surface    corrosion    product",D1),
    aim("thermal    fatigue"),
    ask("corrosion"," is ","inversely   proportional   to   the   crack
            depth",D2), !,
    rule2(Z0), rule21(Z1), rule22(Z2), rule23(Z3),
    append(Z0,Z1,T0), append(Z2,Z3,T1), append(T0,T1,Z),
    append(Z,D,L), append(L,D1,L1),
            append(L1,D2,Conclusion_list),
    asserta(is_a(Fracture,"thermal    fatigue",Conclusion_list)),
    output(Fracture,"thermal    fatigue").
```

```prolog
is(Fracture,"fatigue    at   elevated   temperatures",Conclusion_list):-
     is_a(Fracture,"normal    fatigue",D),
     not(has("fracture","a    surface    corrosion    product",_)),
     aim("fatigue    at   elevated    temperatures"),
     ask("fracture"," has ","a    multiplicity   of   creep   voids   nearby
          it",D1),  !,
     rule3(Z0),     rule31(Z1),     rule32(Z2),
     append(Z0,Z1,T),     append(T,Z2,Z),
     append(Z,D,L),     append(L,D1,Conclusion_list),
     asserta(is_a(Fracture,"fatigue    at    elevated
     temperatures",Conclusion_list)),
     output(Fracture,"fatigue    at   elevated    temperatures").

is(Fracture,"normal    fatigue",Conclusion_list):-
     is_a(Fracture,"normal    fatigue",D),  !,
     aim("normal    fatigue"),
     rule4(Z0),     rule41(Z1),
     append(Z0,Z1,Z),     append(Z,D,Conclusion_list),
     retract(is_a(_,"normal    fatigue",_)),
     asserta(is_a(Fracture,"normal    fatigue",Conclusion_list)),
     output(Fracture,"normal    fatigue").

is(Fracture,"stress-corrosion    cracking",Conclusion_list):-
     not(is_a("fracture","normal    fatigue",_)),
     aim("stress-corrosion    cracking"),
     ask("fracture"," has ","flat   or   grain   facets",A),  !,
     ask("fracture"," has ","corrosion   products   on   its   surface",A1),
     !,
     rule5(Z0),     rule51(Z1),     rule52(Z2),
     append(Z0,Z1,T),     append(T,Z2,Z),
     append(Z,A,H),     append(H,A1,Conclusion_list),
     asserta(is_a(Fracture,"stress-corrosion    cracking",
          Conclusion_list)),
     output(Fracture,"stress-corrosion    cracking").

is(Fracture,"liquid-metal    embrittlement",Conclusion_list):-
     not(is_a("fracture","normal    fatigue",_)),
     has("fracture","flat   or   grain   facets",A),
     aim("liquid-metal    embrittlement"),
     ask("specimen"," is ","contact   with   a   liquid   metal",A1),  !,
     rule6(Z0),     rule61(Z1),     rule62(Z2),
     append(Z0,Z1,T),     append(T,Z2,Z),
     append(Z,A,H),     append(H,A1,Conclusion_list),
     asserta(is_a(Fracture,"liquid-metal    embrittlement",
          Conclusion_list)),
     output(Fracture,"liquid-metal    embrittlement").

is(Fracture,"hydrogen    embrittlement",Conclusion_list):-
     not(is_a("fracture","normal    fatigue",_)),
     not(has("fracture","flat   or   grain   facets",_)),
     aim("hydrogen    embrittlement"),
     ask("fracture"," has ","flakes   or   fisheyes   on   its   surface",A),
     !,   rule7(Z0),     rule71(Z1),
     append(Z0,Z1,Z),     append(Z,A,Conclusion_list),
     asserta(is_a(Fracture,"hydrogen    embrittlement",
          Conclusion_list)),
     output(Fracture,"hydrogen    embrittlement").

is(Fracture,"stress-rupture    failure",Conclusion_list):-
```

```prolog
                not(is_a("fracture","normal    fatigue",_)),
                not(has("fracture","flat   or   grain   facets",_)),
                aim("stress-rupture   failure"),
                ask("fracture"," has  ","a multiplicity of creep voids
                    nearby",A),
                !,   rule8(Z0),   rule81(Z1),   rule82(Z2),
                append(Z0,Z1,T),    append(T,Z2,Z),
                append(Z,A,Conclusion_list),
                asserta(is_a(Fracture,"stress-rupture   failure",
                    Conclusion_list)),
                output(Fracture,"stress-rupture   failure").
```

/*The following are the rules used to determine possible fracture mechanisms.*/

```prolog
rule1(["RULE   1:    The  fracture  mechanism  is  corrosion  fatigue  if  it  is"]).
rule11(["                   fatigue, it has a multiplicity of crack origins, there"]).
rule12(["                 is a pretting product on the surface and the pretting"].
rule13(["                   product is filling the multiplicity of the cracks."]).
rule2(["RULE   2:    The  fracture  mechanism  is  thermal  fatigue  if  it  is"]).
rule21(["                   fatigue, there is a corrosion product on its surface,"]).
rule22(["                 the amount of corrosion along the surface of a crack"]).
rule23(["                 is inversely proportional to the depth of the crack."]).
rule3(["RULE   3:    The  fracture  is  fatigue  at  elevated  temperatures  if  it  is"]).
rule31(["                   fatigue and it has a multiplicity of creep voids"]).
rule32(["                 adjacent to the main fracture."]).
rule4(["RULE   4:    The  fracture  mechanism  is  normal  fatigue  if  it  has  "]).
rule41(["             beach marks."]).
rule5(["RULE   5:    The  fracture  mechanism  is  stress-corrosion  cracking"].
rule51(["                 if it has flat or grain facets and has corrosion']).
rule52(["                 products on its surface."]).
rule6(["RULE   6:    The  fracture  mechanism  is  liquid-metal  embrittlement"]).
rule61(["                 if it has flate or grain facets and the specimen is"]).
rule62(["                 contact with a liquid metal."]).
rule7(["RULE   7:    The  fracture  mechanism  is  hydrogen  embrittlement  if"]).
rule71(["                 it has flakes or fisheyes."]).
rule8(["RULE   8:  The  fracture  mechanism  is  stress-rupture  failure  if  it"]).
rule81(["                 has a multiplicity of creep voids adjacent to the main"]).
rule82(["             fracture."]).
```

```
/ *                        INTERFACE  UNIT                              * /
```

/*The rules used in the INTERFACE UNIT support the user interface. It queries "yes-no" questions" and explains how a conclusion has been reached.*/

/* The **ask** predicate is asking the user if the fracture has specific features.*/

```prolog
        ask(Fracture,Verb,Feature,L):-
            question(Fracture,Verb,Feature,L).
```

/*The following prints the question on the screen, then waits for an **answer**.*/

```prolog
        question(Fracture,Word,Feature,L):-
            word(Word,Verb,Question_word),       blank,
            write(Question_word,"the   ",Fracture,Verb,Feature,"  ?  "),
            readln(Answer),
            respond(Answer,Fracture,Word,Feature,L).
```

/*This following responds to the yes or no answer provided by the user.*/

```prolog
respond("yes",Fracture,Verb,Feature,[S]):-
      concat("FACT:  The  ",Fracture,S1),
      concat(Verb,Feature,S2),
      concat(S1,S2,S3),      concat(S3,".",S),
      asserta(has(Fracture,Feature,[S])).
respond("no",_,_,_,_):-      fail.
respond(Answer,Fracture,Verb,Feature,L):-
      not(member(Answer,["yes","no"])),  nl,      blank,
      write("Input  should  be  either  yes  or  no,  please  retry!"),  !,
      nl,  nl,      ask(Fracture,Verb,Feature,L).
```

/* Here we ask if the user wants to know how the conclusion has been reached.*/

```prolog
ask_how(Fracture,Mechanism):-      nl,      blank,
      write("Are  you  interested  in  how  the  conclusion  was  reached
            ?"),      readln(Answer),  !,  nl,
      explain_how(Answer,Fracture,Mechanism).
```

/* Here the program responds to the users wish to know how the conclusion was reached.*/

```prolog
explain_how(yes,Fracture,Mechanism):-
      how(Fracture,Mechanism).
explain_how(no,_,_).
explain_how(A,Fracture,Mechanism):-
      not(member(A,["yes","no"])),  nl,      blank,
      write("Input  should  be  either  yes  or  no,  please  retry!"),  !,
      ask_how(Fracture,Mechanism).
```

/*Explaining the reasons for the conclusion that has been reached.*/

```prolog
how(Fracture,Mechanism):-
      is_a(Fracture,Mechanism,[H|T]),  !,      blank,      blank,
      write("EXPLANATION:"),  nl,      blank,
      write(H),  !,  nl,
      sub_how(T),  !.
```

/* Displays all the elements the **Conclusion_list** recursively. Lists all the facts and rules
that led to the conclusion.*/

```prolog
sub_how([ ]):-      blank,      blank,
      write("END  OF  EXPLANATION"),  nl,  !.
sub_how([H|T]):-      blank,
      write(H),  !,  nl,
      sub_how(T),  !.
```

```prolog
/*                    SUPPORT  UNIT                              */
```

/*The rules in **SUPPORT   UNIT** give auxiliary features being used in the program.*/

```prolog
word(' has "," have  ","Does ").
word(' is ","  ","Is ").
```

/* Displaying the goal on the screen. */

```prolog
aim(X):-      blank,
      write("GOAL : To  determine  if  the  fracture  is  ",X,".\n"),
      asserta(pass("yes")).
```

/* Displaying the conclusion and asking if the user is interested in how the conclusion has reached */

```
output(fracture,Conclusion):-        blank,
        write("CONCLUSION : The fracture is ",Conclusion,  "."),
        nl,  !,     ask_how(fracture,Conclusion).
```

/* Clearing the database. */

```
clrdbase:-
        has(_,_,_),       retract(has(_,_,_)),       clrdbase.
clrdbase:-
        is_a(_,_,_),       retract(is_a(_,_,_)),   clrdbase.
clrdbase:-
        pass(_),    retract(pass(_)),       clrdbase.
clrdbase.
```

/* Definition of member. */

```
member(X,[X|_]).
member(X,[_|L]):-     member(X,L).
```

/* Definition of append. */

```
append([  ],L,L).
append([Z|L1],L2,[Z|L3]:-     append(L1,L2,L3).
```

/*   END   */

AI APPLICATIONS - DURABILITY/LIFE PREDICTION

OF COMPOSITE MATERIALS

Gary L. Hagnauer and Suzanne G.W. Dunn

U.S. Army Materials Technology Laboratory
Polymer Research Branch
Watertown, Massachusetts 02172-0001

AI technologies are being employed to automate procedures for determining properties and evaluating the environmental aging behavior and durability of polymer composite materials. Knowledge management and the development of an intelligent database management system for integrating materials testing are discussed. The design and operation of "intelligent" laboratory robot work cells utilizing AI techniques to facilitate real-time monitoring, control and integration of robots; planning and scheduling tests; and automating knowledge acquisition are described.

Expert System Applications in
Materials Processing and Manufacturing
Edited by M.Y. Demeri
The Minerals, Metals & Materials Society, 1989

<u>**Introduction**</u>

Assessment of polymer matrix composite material durability and life cycle behavior is a complex, technically demanding and time consuming process that requires extensive testing and different levels of expertise in diverse technical areas (e.g., chemistry, physics, polymer science, chemical engineering and mechanical engineering). Evaluation is further complicated by the inherent complexity of the materials - their composition, morphology, processing and properties are interrelated and susceptible to change. In addition, test results are sometimes difficult to interpret and predictive models are limited. Consequently, the evaluation of polymer composite materials is often a costly and chaotic process - after spending 5 or more years of effort to develop, specify and qualify a material for a particular application, a manufacturer may still not have sufficient understanding of the material for adequate quality control and life cycle management. Considering the many types and applications of composite materials with new materials continually being developed to serve ever more demanding application areas, their efficient and accurate assessment is a matter of paramount technical and economic importance. Today's technology, as implemented, does not result in the effective evaluation of polymer composite materials. Innovative approaches, such as offered by recent developments in artificial intelligence (AI) and laboratory robotics technology, are needed.

Our laboratory is exploring ways to harness AI and robotics technology to automate the testing and evaluation of polymer materials and help guide their specification, design and manufacture.[1] Initially, efforts focused on employing robotics and automated testing technology to increase laboratory productivity and improve the quality of test information. However, our ultimate objective is to integrate and automate, as fully as possible, all processes involved in polymer materials evaluation. Conceptually, our approach is illustrated in Figure 1.

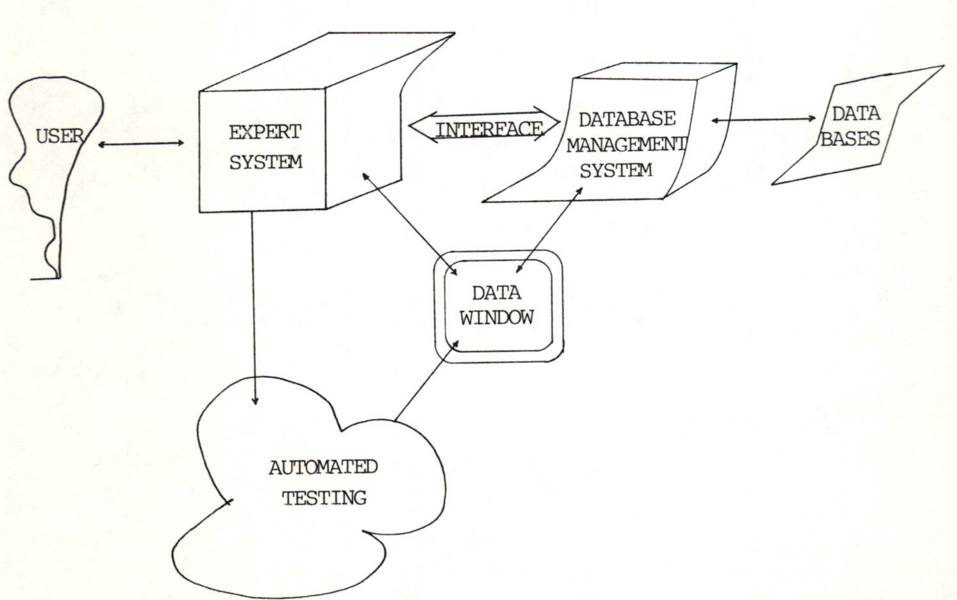

Figure 1 - AI Integration of Materials Evaluation

Two independent systems - an Expert System (ES) and a Database Management System (DEMS) - are coupled. Users communicate through the ES interface with queries and commands to establish "goals" or obtain knowledge. Users also provide the system with expert knowledge and facts and may introduce specimens for automated testing. The ES develops test plans and handles high level test control and decision making functions. AI technologies are also employed in dedicated test methods (e.g., vision - digital image analysis), for robots and sensor integration, and in low level monitoring and control of automated test operations. Test data and related information obtained from test monitors and databases are continuously updated. A Data Window stores and manipulates dynamic information for use by the ES.

This paper describes novel research involving applications of AI technology to automate procedures for determining properties and evaluating the durability of polymer matrix composite materials. Knowledge management and the development of an Intelligent Database Management System will be addressed. The design and operation of "intelligent" laboratory robot work cells utilizing AI techniques to facilitate real-time monitoring, control and integration of work cells; planning and scheduling of tests; acquiring and handling data; and automating knowledge acquisition for environmental aging and durability studies will be described. AI applications for experimental design, robot integration, physical/mechanical testing, image analysis and data interpretation will be discussed.

Knowledge and Database Management Systems

A robust ES and the design and development of an Intelligent Database Management System (IDBMS) are essential for integrating and fully automating polymer materials evaluation. Utilizing commerically available computer hardware and AI/database software tools, our approach in designing an IDBMS is to integrate two distinct systems: an expert system shell-based knowledge management system and a relational database management system (Figure 2). The two systems are "loosely coupled" in the sense that minimal modifica- tions (if any) to either system are necessary with the DBMS acting as a back-end server to the ES, supplying on demand the data that the ES requires.[2] The key issue is the interface between the two independent systems which will enable them to communicate while allowing each to retain its identity to do what it is designed to do best; i.e., the ES is devoted primarily to deductive functions and the DBMS manages the database.

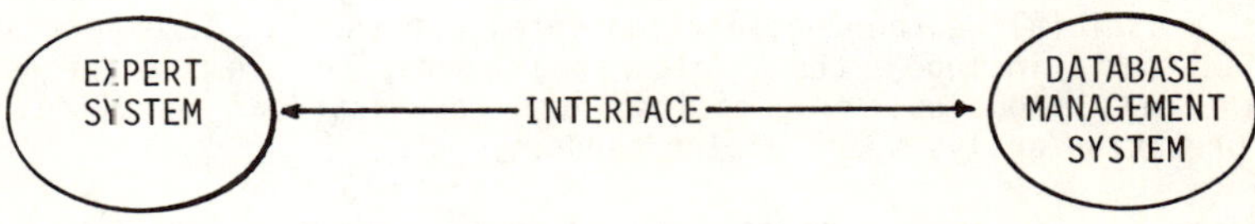

Figure 2 - A Loosely Coupled ES/DBMS

The strategy for a loosely coupled IDBMS has two distinct phases: a computa- tional phase where knowledge generates the queries for the DBMS and an executional phase where queries are processed and results are returned to the ES. Operationally, the ES has a "window" on facts and can access directly only those facts currently loaded on the window. After data "displayed' in the window have been processed, the ES may ask the DBMS for

new data (a process termed as caching). One of the major advantages to this
strategy is the possibility of using existing databases to which the ES can
be connected as one of the application programs. The ES makes calls on
database services in the same way as any other DBMS user. In comparison to
other approaches, a loosely coupled system is more advantageous because the
DBMS can function as a conventional database while at the same time
fulfilling the unique role of a "living database" for automated testing
operations.

To faciliate interfacing the loosely coupled IDBMS, the ES will be implemented using a logic programming language (e.g., PROLOG) which has a common
theoretical foundation with relational databases. Currently, two different
interfacing stategies are being considered: (i) the interface translates an
ES request into the appropriate language[3] and (ii) the database language
is embedded into the PROLOG environment; e.g., PROSQL[4]. Although work has
been done in coupling DBMS's with PROLOG, there is also precedent for
coupling a relational database with a frame-based language, such as KEE.[5]
In the second interfacing strategy, class hierarchy is maintained in the
knowledge engineering environment while the instance frames are maintained
in the relational DBMS.

Conceptually, the ES/IDBMS will operate according to the following processes
or stages:

(1) Document source, treatment, specifications and intended application of
    material: a consultative expert system assists the user in inserting
    quantitative and qualitative information into a material knowledge base.

(2) Acquire relevant supplementary information (where feasible): a consulta-
    tive expert system/database recommends appropriate tests, characteriza-
    tion procedures and analyses. User is referred to pertinent property
    data and/or general relationships for similar materials to estimate and
    rate important properties. Possible problem areas and critical factors
    are identified and test plans are recommended based on requirements,
    resources and urgency.

(3) Prepare and dispense test specimens: a consultative expert system
    examines variables (composition, processing, conditioning, post-
    treatment) and recommends type, number and distribution of test
    specimens.

(4) Automated testing: AI/robotics technology is employed for specific
    property evaluation and more extensive durability testing (accelerated
    aging, artificial weathering, environmental exposure). AI/expert system
    applications include test planning, decision making (conflict
    resolution), robotics, diagnostics, data acquisition/ handling, data
    interpretation/analysis and vision/sensors.

(5) Assessment: AI/expert systems and advanced computer hardware assist
    materials evaluation in areas of knowledge base generation, machine
    learning (physical law discovery), life cycle prediction (simulation/
    modeling), material performance rating and communications/reporting.

Automated Testing

AI techniques are employed with robotics and test instrumentation to develop
"intelligent" laboratory robot work stations for automated testing. Work
cells are designed to compensate for the lack of technicians and yet provide
needed flexibility for the automation of tests that are repetitive, tedious,

operator-sensitive and/or hazardous. The laboratory robots are rather primitive systems designed to perform "unit operations" (e.g., picking and placing specimens, transferring and mixing liquids, and weighing samples) in highly structured "cells".[6] The tests are quite involved and require not only a high degree of manual dexterity but also flexibility in sequencing and integrating operations. Our approach has been to design and engineer robotic work stations to operate as either stand-alone or integrated materials testing systems and then provide the robots with sensing and cognitive abilities to enable them, at least partially, to determine their own actions. AI techniques facilitate real-time monitoring, control and integrated operation of the robotic work cells and help alleviate many problems associated with data handling, communications and reporting.

Progress in developing and integrating intelligent robot work stations has been both incremental and evolutionary. Each adaptation of robotics has involved design, demonstration, implementation, and redesign or extension. As each robot work cell is extended and eventually integrated with other work cells/operations, a more versatile system is generated which represents a further phase of development. Each phase of development also increases the complexity of the work cell system and requires higher level AI/robotic functions.

Immersion Testing

The system for automated immersion testing of composite test specimens is typical of fully automated robotic work cells being developed for materials evaluation.[7] The test involves measuring weight changes of thin (<3 mm, thickness) specimens as a function of the time the materials are immersed in a liquid at a specified temperature. Depending upon a specimen's chemical composition and physical form, a test may run for several months and generate large amounts of data that must be collected and evaluated. When run manually, immersion testing is tedious and time consuming. Measurements are limited to an eight hour work schedule and subject to human errors. Also, manpower constraints limit the number of samples and the variety of conditions under which materials can be tested. When water is used as the immersion liquid, the test is essentially the ASTM Standard Test Method (D570) for Water Absorption of Plastics.[8] Determinations of the relative rate of water absorption are important in evaluating the effects of moisture exposure on such properties as mechanical strength, electrical resistivity, dielectric losses, physical dimensions and appearance. Moisture content has significant effects on physical-mechanical properties and the long term durability of polymer matrix composite materials. To assess moisture sorption/diffusion behavior, it is imperative that immersion tests be run in a consistent manner on uniformly shaped specimens.

The work cell includes six major pieces of equipment - a Zymate robot arm and controller (Zymark Corp., Hopkinton, MA); a Zymark model Z410 capping station; a top-loading, dual-range analytical balance (Mettler A163); a specimen blotting station; a thermostated oven for holding specimen jars; and an IBM PC/AT system. Except during periods of weighing, each specimen (typically, 25 mm x 25 mm x 2 mm) is immersed in the test liquid and maintained at a constant temperature inside a specially designed screw-cap glass jar (100 mL). Robot operations include capping and uncapping specimen jars, removing specimens from their jars, blotting and weighing specimens, and replacing specimens in their respective liquid environments. Optical sensors and relay switches are used throughout the test routine for verification. For all possible problems, as well as the sequence in which they occur, the robot must recognize that there is a problem, define the problem, decide how best to resolve the problem, perform the necessary operations to overcome the problem, and enable the system to resume testing.

This is an AI application area and a critical feature, especially since the system operates unattended and measurements are taken overnight and during weekends.

Overall control of the "intelligent" immersion testing work cell is directed by an IBM PC/AT computer using the DOS operating system and a menu-driven program. A front-end expert system shell was designed to assist users in introducing new specimens, planning tests and developing the knowledge base. Special AI routines (quasi production rule based expert systems) provides needed flexibility for the testing procedure and real-time control (decision making) in running tests. The IBM computer handles high level decision making (e.g., scheduling of weighings, checking whether a test is running properly, deciding whether a test should be terminated, and resolving anomalies) and directs the Robot Controller. The Controller directs the robot and executes routines dictated by sensor responses and low-level decision making programs written in a special programming language. The IBM PC/AT maintains a database and graphical results are automatically created and displayed using a procedure written in the RS1 (a DOS graphics package) programming language.

The intelligent robotic work cell significantly reduces the cost of perform-ing immersion tests while increasing productivity 10-fold. Automation improves both the precision and accuracy of the test by (i) eliminating human operator errors, (ii) providing for more frequent data acquisition, particularly at critical times (beyond normal working hours), (iii) stan-dardizing operations (measurements are made exactly the same way each time), and (iv) requiring uniform specimens and better sampling (e.g., upright test specimens in the immersion liquid). The application of robot sensors and advanced programming techniques for planning/scheduling weight measurements, resolving conflicts (about 80% of the software is devoted to error detection and correction routines), and data acquisition and assessment ultimately led to the system's high degree of reliability and efficiency.

Durability Testing System

The immersion testing work cell has been extended and currently is being incorporated with automated mechanical testing and dynamic mechanical analysis work cells to automate durability testing of polymer composite materials. The scheme for the automated durability testing system is illustrated in Figure 3. (Dashed lines indicate planned work cells/test stations) The durability test system is designed to handle large numbers of specimens and receive specimens either individually or in racks from weathering chambers and environmental exposure sites. The basic immersion testing work cell now includes three (3) robots which share tasks and must have their operations coordinated to prevent conflicts and achieve optimum efficiency. Aging ovens, an additional analytical balance and a blotting station have been added to the system. Further enhancements include an automated micrometer to measure specimen thickness, a liquid level sensor and delivery station, drying and vacuum ovens to remove absorbed moisture and condition specimens for physical and mechanical testing, robot end-effectors and tactile sensors, customized specimen holders, and special devices for transferring specimens between work stations. Also, the specimen jars are modified to hold a wider range of specimen sizes and shapes to accommodate various test requirements.

The mechanical testing work cell is fully automated. Individual specimens or racks of specimens are transferred from the conditioning station to the mechanical testing work cell which includes an Instron Universal Testing Instrument (Instron Corp., Canton, MA), a Zymate robot arm and controller, specimen racks, a bar code reader, and an IBM PC/AT. The Instron Tester is

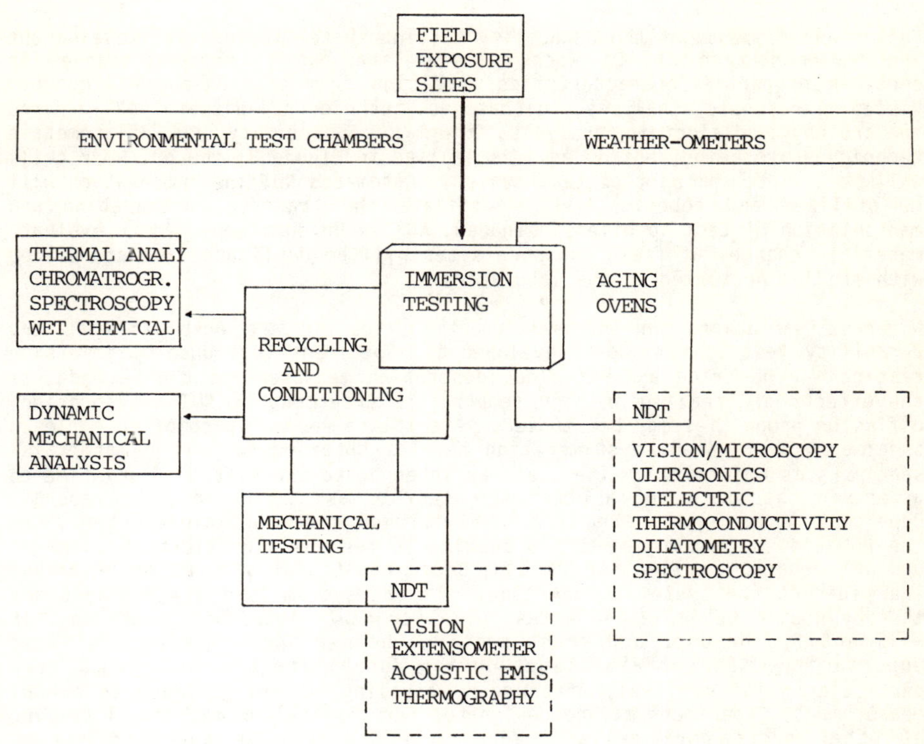

Figure 3 - Scheme for Automated Durability Testing System

a heavy duty instrument with hydraulic grippers and load cells selected especially for tensile and flexual testing of polymer matrix composite materials. Currently, the robot merely transfers test specimens from specimen racks to the test apparatus and then removes specimens from the apparatus after the test is completed. However in the near future, a vision system will be added to the work cell to assist the robot in positioning and removing specimens before and after testing and to provide additional information about the quality and interpretation of test results. Preliminary work in our laboratory has shown that a vision system, when properly implemented with digital image analysis, is a powerful tool for identifying specimen fracture behavior and monitoring changes in the surface characteristics of specimens due to weathering. Other non-destructive tests (NDT) under consideration include a non-contact extensometer for measuring specimen strain during tensile testing, infrared thermography to measure specimen defects and heat generated during fatigue cycling tests, and acoustic emission for monitoring damage to specimens incurred during mechanical testing.

The dynamic mechanical analysis (DMA) work cell includes two DuPont Model 983 automated DMA instruments, specimen racks, and an IBM PC. The work cell is only partly automated in that a robot delivers specimens for analysis, but a technician must load specimens into the instruments for testing. Similarly, various thermal, chromatographic, spectroscopic and wet chemical analytical techniques are automated, but human intervention is required to facilitate testing. Future work involves fuller integration of DMA and the associated analytical techniques into the durability test system.

Future enhancements of the durability testing system include the development and implementation of NDT work cells or stations to measure changes in specimen properties/characteristics resulting from environmental exposure. Vision/microscopic analysis, ultrasonic techniques, dispersive infrared spectroscopy, dielectric analysis, thermal conductivity and dilatometric techniques are being considered. As implied in Figure 3, the NDT work cells will be an integral part of test system. Automated NDT instrumentation will be utilized and robotics will facilitate the transfer, preparation and manipulation of test specimens. Indeed, NDT techniques employed to evaluate materials changes at field exposure sites will be duplicated and correlated with studies performed in the laboratory.

Numerous variations in specimen treatment/testing are possible with the durability test system under development. Specimens that undergo immersion testing may be dried and recycled, desorption measurements can be made, or the effects of freezing or high temperature treatment on NDT and sorption/ diffusion properties may be studied. A complete record is compiled for each specimen and a database summerizing results obtained for other materials/ specimens is maintained. The user may interrogate the system at any time to ascertain the status of a particular test or ask for a progress reports. Depending upon relative changes in measurements and anticipated behavior, the robotics system may modify a specimen's test plan or alert the user of unusual behavior. The user can introduce additional samples or interrupt and redirect the system at any time. If the system is fully employed and thereby unable to implement a test plan for newly introduced specimens, it will notify the user and start testing the new specimens at the first opportunity. At certain stages during durability testing (e.g., at a particular moisture level, after each recycling, or in response to an NDT measurement), specimens may be designated for physical or mechanical testing at other robot work cells. When this occurs, the specimen may be reconditioned and transferred to the designate work cell for analysis or further testing (e.g., chemical analysis or mechanical testing).

AI Integration of Materials Evaluation

Computer and software requirements of the stand-alone work stations are already demanding and becoming increasingly so as tests become more complex. Although the robot work cells are individually programmed to recognize problems and resolve conflicts, full integration of the various automated elements presents special problems - choice of architecture and interfacing techniques and coordination of all elements of all functions involved - where conventional approaches (e.g., handshaking) for control and monitoring may not provide the best solution.

Implementation of AI technology is key to the successful application of robotics for materials evaluation. Robot systems must be equipped with some intelligence for the various activities to be coordinated and for the components to be utilized in the most efficient and productive manner. Thus far, we have employed advanced computer programming, but rather primitive AI techniques, to make the robots behave intelligently in a highly structured domain - modular programming (microcomputer directing the robot controller), concurrent operations (multiple arms and end-effectors), problem solving (conflict resolution in real time), sensor integration (optical and mechanical), and expert systems (planning, database, interpretive). Improved hierarchical control and an expert robot task planner (using a knowledge base for sequence and time planning per specification of goals) need to be developed. Vision and other forms of potentially intelligent sensors need to be implemented to help coordinate activities and make both qualitative and quantitative improvements in tests. AI techniques are needed not only to manage testing and handle information, but also to interpret data (discover relationships), develop better models for estimating material lifetimes, and guide users in decision making. For example, once sufficient experimental data is obtained, heuristic techniques could be applied to guide the inductive process of determining empirical relationships by massaging data and searching for correlations or possible functional relationships.

References

1. G.L. Hagnauer and S.G.W. Dunn, "AI Methods for Evaluation of Plastics and Composite Materials", _Artificial Intelligence Applications in Materials Science_ (Warrendale, PA: The Metallurgical Society, 1987) 157-73.

2. B. Napneys and D. Herkimer, "A Look at Loosely-Coupled Prolog Database Systems", _Expert Database Systems_ (Proceedings from Second International Conference), ed. L. Kershburg, Benjamin-Cummings Publishing Company: Menlo Park, CA, 1989.

3. S. Ceri, G. Gottlob and G. Wiederhold, "Efficient Database Access from Prolog", _IEEE Transactions on Software Engineering_, 15 (2), (1989).

4. C.L. Chang and A. Walker, "PROSQL: A Prolog Programming Interface with SQL/DS", _Expert Database Systems_ (Proceedings from First International Workshop), ed. L. Kershburg, Benjamin-Cummings Publishing Company: Menlo Park, CA, 1986.

5. R.M. Abarbanel and M.D. Williams, "A Relational Representation for Knowledge Bases", _Expert Database Systems_ (Proceedings from First International Conference), ed. L. Kershburg, Benjamin-Cummings Publishing Company: Menlo Park, CA, 1987.

6. J.D. Kleinmeyer, "Requirements and Selection of Laboratory Robotic Systems", MTL Technical Report TR 89-57, U.S. Army Materials Technology Laboratory: Watertown, MA (July 1989).

7. S.W.G. Dunn, "Automated Immersion Robotics System Hardware and Software Design", MTL Technical Report TR 89-36, U.S. Army Materials Technology Laboratory: Watertown, MA (May 1989).

8. ASTM D570 in _Annual Book of ASTM Standards, 08.01_ (Philadelphia, PA: American Society for Testing and Materials, 1987).

USING AN EXPERT SYSTEM TO INTERPRET FINITE

ELEMENT ANALYSIS PROCESS HISTORIES

K. J. Meltsner

Materials Research Center
GE Corporate Research & Development Center
Schenectady, NY 12301

Abstract

Finite element analysis (FEA) is an essential part of many manufacturing design systems. However, FEA programs produce large data files which must be interpreted by the design engineer. As part of a manufacturing design workstation, a metallurgical expert system has been provided to interpret FEA simulations of Ti-6Al-4V thermomechanical joining. The expert system includes a history scanner to reduce the FEA data files and a metallurgical knowledgebase to predict the part's microstructures and properties. The program also includes a library of micrographs and an explanation module which are available to assist the user in understanding its predictions. The current system is limited to the analysis of a particular part, but its methods are applicable to other materials and manufacturing methods.

Expert System Applications in
Materials Processing and Manufacturing
Edited by M.Y. Demeri
The Minerals, Metals & Materials Society, 1989

Introduction

It is now possible to use finite element analysis (FEA) to model many thermomechanical processes, but FEA does not provide predictions of quantities important to metallurgists such as sizes and shapes of microstructural features or the presence of specific phases, or predictions important to designers such as the piece's mechanical properties. This paper will discuss the use of expert systems to provide interpretations of FEA results.

With the availability of powerful expert system shells, the primary problem becomes the organization of a body of metallurgical knowledge into a form suitable for expert system use. Our approach has been to relate all of the processing effects and properties predictions by correlating both with the possible microstructures. This approach is suitable when the process by which the material's microstructure develops is well-understood.

The expert system described in this paper is a diagnostic aid and post-processor for an FEA system(1) which provides an analysis of the deformations and temperatures during upset joining of a Ti-6Al-4V part. It is used by the part's designer to predict whether undesirable metallurgical conditions will occur during the manufacturing process. The system is not tightly coupled to the FEA program, but interprets its output files to provide these warnings and explanations of possible problems.

Selecting critical information from process history files

Simple quantities

The scanner routine can find many simple quantities without additional programming. These include initial and final values, as well as maximum and minimum values. More than one type of quantity can be found for a process variable. For example, it might be useful to find both the initial and maximum temperatures during a process. The ES designer can also specify units conversion routines to change process history values into the expert system's internal units.

After the scanning is completed, the quantities are available for use by the rest of the expert system. The system creates an instance to hold the values of the scanned quantities for each node, which allows the shell's inference engine to access them.

"Special" quantities

It may be necessary to find the value of a quantity which is not available by simply scanning the process history. These quantities are known as "special quantities" and are calculated during the scanning process. Special quantities give the scanner much of its power and flexibility; they may be used, for example, to automatically select starting microstructures for a piece or to calculate cooling rates from transition temperatures.

The scanning routine actuates the special routines to calculate such quantities from one or more of the process variables. A routine can access any of the quantity values present in the process history or its own local variables. At the end of scanning process, the routine's results are used in the same way as simple quantities.

The expert system includes a library to help the programmer to create special routines. A simple LISP form or equation can be written to calculate the quantity and the system will automatically generate a routine to calculate this quantity during the scanning process.

Creating the knowledgebase for a metallurgical expert system

Knowledge sources

During the development of our metallurgical expert system, several important limitations of the expert system approach became clear. Traditionally, expert systems involve an expert who knows all there is to know about a subject and a knowledge engineer who interviews the expert and organizes the knowledge into a form the computer can use(2). Unfortunately, for many metallurgical problems, there is no omniscient expert, and it becomes necessary to use a variety of knowledge sources.

The best sources for information are the handbooks and critical reviews about the material and process. Most include information on the effects of heat treatments and mechanical treatments, typical mechanical properties, etc. If a material or process is not common enough to be described in the open literature, the problem is more difficult. In this case, local experts have to be convinced to organize their experience into the equivalent of a review article.

This process usually shows that there are significant gaps in the available knowledge. Even well-understood materials have such gaps, because most of them are used with a limited range of possible thermomechanical treatments. If more knowledge is needed, the expert must design experiments that will show the effect of more general thermal and mechanical treatments upon the material's structure and properties.

Expert systems do not provide a "magic" way to predict the performance of materials or processes. In fact, since an expert system must understand the entire range of possible process conditions, it requires substantial expertise on the material's structures and properties. Normally, unusual situations can be explored on a case-by-case basis by human experts, with additional tests run as needed. Unfortunately, expert systems tend to be "brittle" when they run up against such cases; they rarely have the facilities to handle unusual situations. The use of a general microstructural hierarchy can help the expert system make conservative guesses, but until more sophisticated simulation systems can calculate material properties from first principles, expert systems will only be as good as their knowledgebases.

In fact, expert systems should not be viewed as a way to replace experts, but as way to make their expertise more widely available. Requirements and production conditions change, which means that the knowledgebases must change as well. Expert systems are essentially automated handbooks capable of applying complex rules and performing detailed calculations, but inherently limited by the creators' knowledge. Expert systems also require the same attention as handbooks to such issues as updates and user skills.

Representing microstructures and processes

The essential problem of a metallurgical expert system is the representation of the materials and the processes of interest. There have been a variety of representation schemes at various levels of detail(3, 4); we chose to represent the metal at the level of detail visible with optical microscopy. In addition, we chose to describe these structures as a metallographer would. Certainly, many important microstructural features cannot be represented this way, including any features at an atomic scale. This scheme does represent the relationships, such as phases present or particle shapes and sizes, which are important in our process.

Hierarchical model of possible microstructures

The first step in setting up our metallurgical expert system was to define the range of possible microstructures. Simple problems may require one or two parameters such as grain size and phase fraction, but in more complex problems, a variety of microstructures may be possible. To organize them, a hierarchical approach can be used. This approach describes microstructures as variants on more general structures (Fig. 1). The hierarchical approach allows the system to make reasonable guesses when quantities are not known. Our expert system uses a frame system(5) to describe the set of possible microstructures in terms of the significant structural parameters. To provide more specific information, it also

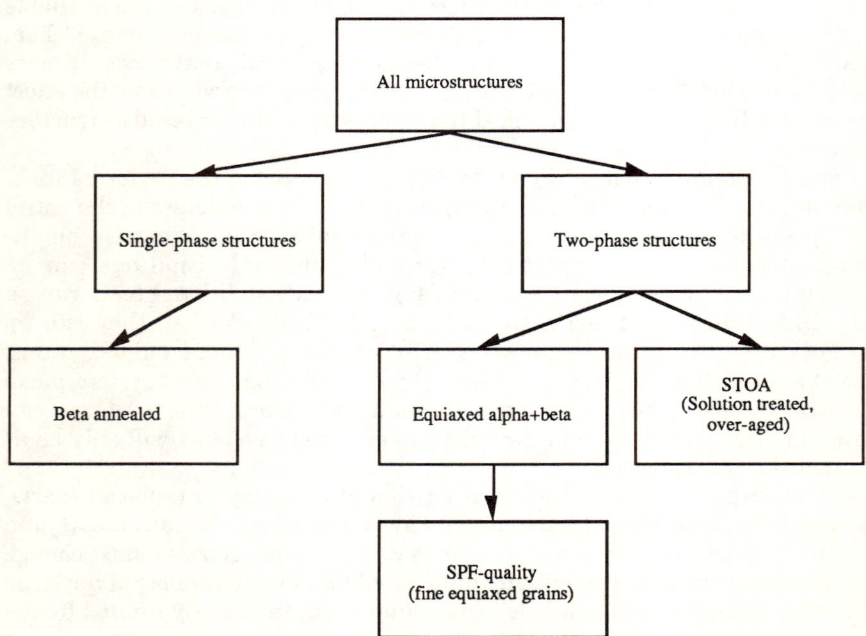

Figure 1:   Sample hierarchy of structures for Ti-6Al-4V.

includes frames that describe each set of possible variants, and frames to describe even more specific variants of those frames. At each level, the designer can specify default values for microstructural parameters. When a property is not known, the frame system looks for the default value specified in the most specific frame. The following is an example from the expert system (edited for clarity) of how the frame system provides increasingly specific information for a family of structures:

```
(define-frame TWO-PHASE
  (:is STRUCTURE
   :doc-string "Two phase structure")
  (grain-size-1)
  (grain-size-2)
  (phase-fraction-1)
  (aspect-ratio-1 :default-value 1)
  (aspect-ratio-2 :default-value 1))

(define-frame ALPHA-BETA
  (:is TWO-PHASE
   :doc-string "alpha/beta titanium")
  (phase-fraction-1 :default-values .85)
  ;typical value for phase fraction(alpha)
  (aspect-ratio-1 :default-value 1)
  (aspect-ratio-2 :default-value 1)
  (constituent-1 :default-value alpha)
  (constituent-2 :default-value beta))
  ;names describing constituents

(define-frame STOA
  (:is ALPHA-BETA
   :doc-string "STOA alpha-beta Ti-64")
  (grain-size-1 :default-value 65.0)
  ; STOA is solution treated near beta transus
  (phase-fraction-1 :default-value .20))
```

To use the information stored in the frames, the system creates *instances*. Each instance is a member of the set described by the frame and holds the information about the structure of one homogeneous area. Since the expert system is used to interpret FEA results, we assume the material in the vicinity of each FEA node experiences the strains and temperatures at that node, allowing us to assume that an instance can describe the microstructure of that material. This is the fundamental link between the FEA nodes and the expert system's predictions.

Organizing processes which affect microstructures

The designer must correlate the processes that occur at a microstructural level with the overall thermomechanical process. This is a task that needs to be performed by an experienced metallurgist. For example, if a piece is held at a temperature in the α/β region, most metallurgists would predict there will be a period of static grain growth and a change in phase fraction of α to the equilibrium amount. In addition, experience with Ti-6Al-4V would allow a metallurgist to predict that a rapid drop in temperature will not substantially affect the phase fraction of α present in the final part(6). The choice of appropriate microstructural processes is a fundamental part of the creation of the knowledgebase. This phase of the expert system design requires a firm understanding of the times, strains, and temperatures required for each possible microstructural process. The scanner supports the expert system designer by providing the ability to find metallurgical quantities of interest and by reducing the huge process history files generated by the FEA code, but it is up to the designer to specify what information will be needed.

Programmed selection from initial microstructures

To accommodate pieces made from non-homogeneous starting material, the system uses a file containing the material present at each node in the piece. The possible materials can be described as specializations of the appropriate micro-structures. For example, a piece of superplastic forming (SPF) grade material has an equiaxed α-β structure, but should also have a grain size of 6-10 μm(7). The hierarchical library of microstructures allows the ES designer to create a frame for all SPF-quality material which inherits all of its properties from the equiaxed α-β frame and which adds the additional property of its fine grain size. The expert system automatically generates an instance of the frame describing the SPF-grade material for each node at which there will be SPF-grade material.

Microstructural prediction rules

The system uses two types of microstructural rules: rules which change one or more quantities in a structure, and rules which change one structure to another. An example of the former would be a grain-growth rule relating a thermal exposure to a change in the material's grain size, and an example of the latter would be a phase transformation rule describing the change in micro-structure if the piece experiences a temperature above a phase transition temperature.

The rules are written to alter the value of the property stored in the microstructural instance. A quantity change can also trigger rules or other actions, such as a diagnostic predicting high scatter in mechanical property measurements if the grain size is too large. This is an example in pseudo-English of a grain growth rule which will change the value of the grain-size property stored in an instance:

```
ALPHA-BETA-GRAIN-GROWTH rule:
    IF    there is a structure which is alpha-beta and
          there is a node for this structure
             which has a time-step of "time" and a
             temperature of "temp"
   THEN the structure's grain size of the alpha phase is
             the value of alpha-beta-grain-growth(temp, time)
```

where *alpha-beta-grain-growth* is a LISP function defined in terms of temperature and time.

The rule system finds all instances of the structures which match the conditions in the first part and then carries out the actions in the second part of the rule upon them.

Rules which change the structure are different. In this case, the rule specifies a different frame to describe the microstructure at the node:

```
BETA-TRANSUS rule:
  IF    there is a node
          with a temperature of "temp" and
        temp exceeds the beta transus temperature
  THEN the structure of the node will be beta.
```

The system includes a small routine (a "daemon") to manage the creation and deletion of the structure instances. This allows the rule creator to specify a new structure without worrying about the mechanics of managing the structure instances.

Properties prediction rules

The current expert system does not attempt to make exact properties predictions. It is used to predict when there will be problems with a part, not how good it will be. The system does not use uncertainty methods (i.e. as a quantity nears a particular value, the predicted probability of problem increases), but instead uses threshold values (i.e. above a certain level, there will be a problem) for its properties predictions. Currently, the system predicts a problem with the piece if any property at any node is deficient. Future plans include a link to a CAD database so the system can find if a node will be in the finished part, since the pieces undergo substantial machining.

User-interface issues

Explaining metallurgical conclusions

The shell used in this project has the ability to maintain justifications for any inferenced conclusion. The expert system uses this capability to provide an explanation facility. The user can select any node on the finite element mesh and the system will explain any conclusions that concern it. The justifications for the conclusions can then be examined (and their justifications as well) all the way back to the information scanned from the process history file. The following example demonstrates how the chain of deductions can be followed back to the quantities scanned from the process history file:

```
Node #158 has the following information:

We predict that:
This node will have unacceptably poor properties.
The fact(s) that lead to this conclusion:
        The node's beta structure will have low Charpy values.

We predict that:
This transformed beta structure will have poor Charpy toughness,
since it has over-elongated beta grains.
The fact(s) that lead to this conclusion:
        The node's beta structure will have over-elongated beta.

We predict that:
The beta grains in a beta structure will have undesirably high
aspect ratios.  (With a "stretch" above 3, the grains will be
highly elongated.).
The fact(s) that lead to this conclusion:
        LAST-STRETCH-1 of the NODE is 3.6505
        STRUCTURE of the NODE is BETA

We derived the prediction of a beta structure from the
information scanned from the process history.
```

User displays

The expert system uses DEACON, a GE proprietary user-interface library. This allows the expert system to look like other GE design programs and reduces the amount of training needed to use the system. DEACON provides several important features, including finite element mesh displays, scalar quantity graphs, menus, and a job control system. Since the expert system is written in LISP and DEACON is written in FORTRAN, a package was written to provide the needed foreign function interfaces.

The shell's frame system provides an object-oriented interface to DEACON. In this way, the expert system can be moved to other environments if necessary. Also included in the library of user-interface objects is an image display object. It is used to display micrographs and other digitized images. Figure 2 shows a typical session.

Providing an on-line metallographic atlas

The expert system provides a simple on-line metallographic atlas. When the user needs a display of a predicted microstructure, the system selects the closest match from its library of digitized images. The selection process uses several rules to select the best match. First, the images with a structure closest in the hierarchy to the desired microstructure are selected. The system then calculates the difference in microstructural parameters from the target structure. This allows the system to rank the images. Finally, the expert system chooses the image with the fewest differences. If there are several images with the same number of differences, the system displays a menu with the differences between the images and allows the user to select which ones to display. If there is no good match, the system tries to find a better match with the parent structure of the target.

Discussion

Importance of metallurgical approach to problem

The expert system requires a metallurgical approach to the post-FEA interpretation. Previous efforts showed that direct correlations of process conditions with final properties were weak. However, metallurgical analyses of the pieces showed significant differences between otherwise identical parts. As a result, the expert system needed to concentrate on metallurgical differences rather than process differences. The system has not been in use long enough to verify its predictions, but there is confidence that its predictions match those of experienced metallurgists.

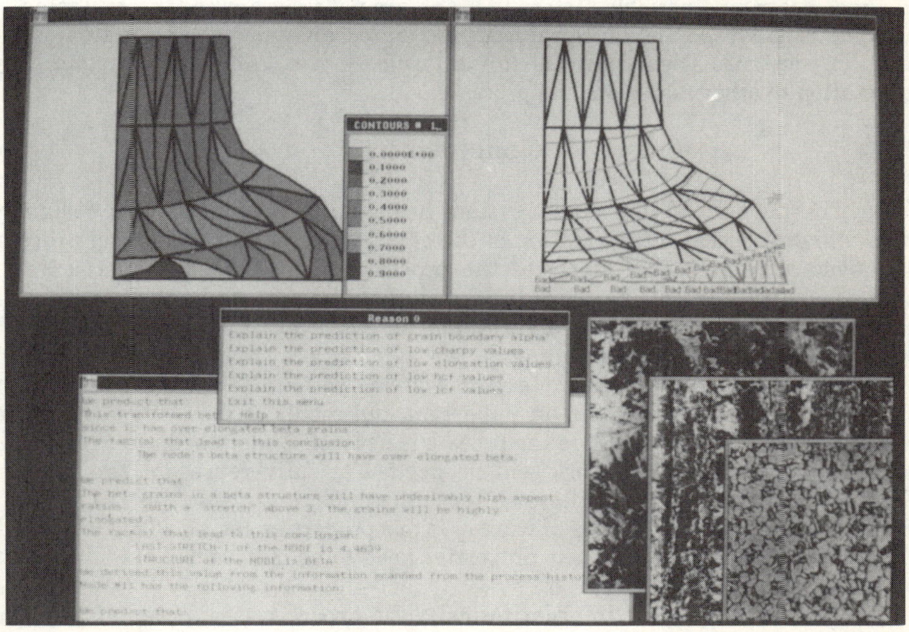

Figure 2:   A session using the metallurgical model expert system. The picture shows microstructural displays (lower left), mesh displays (upper half of screen), a text window, and a menu-driven explanation system.

Targeted user

The expert system has been designed to fit the needs of a non-metallurgist process designer. In addition to the metallurgical constraints, there are many mechanical constraints on the process, including forming press loads, heat transfer during set-up, etc. A conventional finite element code is available to calculate the mechanical aspects of the process. This expert system can then be used to check whether the process design will cause any metallurgical problems. The system provides a complete explanation facility for its predictions which is used to explain the relationship between predicted properties and microstructures. However, the expert system must also be useful for the metallurgists who will check the system's prediction. The system can display both metallurgical predictions (grain sizes, phase fractions) and areas with poor mechanical properties.

The expert system is also designed to be maintained by non-computer programmers. While maintenance will require knowledge of expert system concepts, templates and libraries are available to simplify the task. The frame system simplifies the maintenance and creation of rules, microstructure frames, and special quantities for the scanner. Once the format is learned, additional quantities, microstructures, and diagnostics may be added without changing the rest of the system. In addition, the process history file formats are part of an instance describing possible history files and are not programmed directly into the system. Finally, the metallurgical functions, rules, and microstructure frames are kept in separate files from the underlying system, which should simplify adaptation to other materials and processes.

Conclusions

The purpose of this expert system has been to provide process designers with a metallurgical interpretation of the effects of a thermomechanical process. The system relates the effects of the process and predicted properties at a microstructural level. We chose this level of abstraction because we could find sufficient understanding of the alloy system at this level; at a more detailed level, adequate information about the process would not be easy to obtain.

During the creation of the expert system, we found that the creation of a complete knowledge base was an important end-result of the system design as well. In other situations, where the process is not well-understood, we expect that substantial metallurgical work will be needed to fully represent the system. Surprisingly, many common alloys are not fully understood, since, as long as they are used within a defined process window, their properties can be easily predicted. Difficulties will arise when their properties need to be understood outside of this process window.

We also found that in order to make the expert system acceptable to the end-user designer, its explanations must be fully understandable by both the designer and other metallurgists. The expert system includes a complete explanation facility and an on-line metallographic atlas in addition to displays of important quantities on the FEA mesh.

Finally, we kept the metallurgical and computer science sections of the system separate, which should allow the system to be adapted to similar problems, and provided facilities to minimize the system's maintenance effort. In the future, we hope to provide a core system for processing FEA results for many well-understood materials and manufacturing operations.

System implementation

The expert system requires a Digital Equipment Corporation VAXStation II/GPX workstation. It is written in VAX LISP/VMS, Digital's implementation of Common LISP. A special version of GoldWorks (Gold Hill Computers, Inc.) shell was used as a base for the expert system. DEACON, a GE-proprietary user-interface library, provides its menus and displays.

As of the summer of 1989, the first version of the expert system was completed. It is expected to be installed and in use by the end of 1989.

Acknowledgements

I would like to acknowledge the assistance of Dr. M. F .X. Gigliotti with titanium metallurgy, Dr. H. S. Spacil with the initial design of the metallurgical expert system, Ms. E. M. Perry for integrating the FEA program and the expert system, and Dr. H. A. Nied with the FEA code. In addition, Mr. S. P. Collins wrote several critical programs and libraries for this project.

Bibliography

1.   H. A. Nied, "Interface Displacement Characteristics of Upset Welding," *Second International Conference on Trends in Welding Research*, ed. (Gatlinburg, Tenn.: ASM International, in press).

2.   J. R. Walters and N. R. Nielsen, *Crafting Knowledge-based Systems* (New York: Wiley-Interscience, 1988).

3.   D. E. Marinaro and J. W. J. Morris, "Research Towards an Expert System For Materials Design," *Artificial Intelligence Applications in Materials Science*, ed. R. J. Harrison and L. D. Roth (Orlando, Florida: The Metallurgical Society, 1986), 49-77.

4.   I. Hulthage *et al.*, "The Metallurgical Database of ALADIN—An Alloy Design System," *Artificial Intelligence Applications in Materials Science*, ed. R. J. Harrison and L. D. Roth (Orlando, Florida: The Metallurgical Society. 1986), 105-122.

5.   P. H. Winston and B. K. P. Horn, *LISP* (Reading, Mass.: Addison-Wesley, 1984), 311-319.

6.   A. K. Ghosh and C. H. Hamilton, "Influences of Material Parameters and Microstructure on Superplastic Forming," *Met. Trans. A*, 13A (May 1982), 733-742.

7.   C. H. Hamilton, "Superplasticity in Titanium Alloys," *Superplastic Forming*, ed. S. P. Agrawal (Metals Park, OH: American Society for Metals, 1985), 13-22.

AN EXPERT SYSTEM TO RECOMMEND ROLL ADJUSTMENTS IN A RAIL MILL

M. M. Vyas, D. J. Renn and S. M. Harpster

Bethlehem Steel Corporation
Bethlehem, PA

Abstract

At Bethlehem Steel's Steelton Plant, floormen and rollers are using an expert system that recommends roll adjustments for the 28" rail mill. The objective is to produce rails within strict dimensional tolerances. Based on the rail sample dimensions and other process data, appropriate roll adjustments are made to maintain consistent rail dimensions. A PC-based expert system was developed and installed in the mill to recommend the "best" of several possible complex roll adjustments using real time data input by mill operators. The experience and knowledge of the mill rollers and theoretical rolling knowledge were used to establish the roll adjustment criteria for the expert system. It also includes supporting computer programs for rail dimension trend charting, data reporting and statistical quality control charts.

Expert System Applications in
Materials Processing and Manufacturing
Edited by M.Y. Demeri
The Minerals, Metals & Materials Society, 1989

Introduction

Due to the increased need to minimize hot-rolled section production costs while maintaining high product quality, steel rolling mills are continually looking for ways to improve production methods through standardization of operating procedures and increased reliance on computers in decision making. The operation of a steel rolling mill is usually learned by the operator through experience of working in the mill for many years. Since the guidelines for many complex mill adjustments are developed empirically, they often vary from operator to operator. Upon termination of employment of an expert operator, experience with rolling procedures may also vanish from the scene.

Expert systems technology allows retention of expertise and implementation of standard operating practices because of its ability to incorporate quantitative and qualitative logic and interface with other computer programs such as data trend charting and Statistical Process Control charting. In addition, since it is relatively easy to change the knowledge base, the expert system could be progressively updated as more is learned about the rolling process.

An expert system is a computer program that is used to emulate the decision making process of a human expert in a specific knowledge area of limited scope. Most expert systems consist of a knowledge base, an inference engine, and a user interface. The knowledge base contains the information that the expert system uses to reach a conclusion. A common method for representing knowledge is through the use of rules i.e., "IF (condition...) THEN (conclusion...)". The inference engine determines which rules to process in order to reach a conclusion. An inference engine can use either forward chaining (i.e. data driven) or backward chaining (i.e. goal driven) reasoning. The user interface provides a dialogue with the user to obtain information and display results [1,2].

Hot Rolling of Steel Rails

Hot steel rolling is a deformation process in which a steel workpiece is heated to rolling temperature, typically in excess of 2,000°F, and then rolled between two or more rotating rolls. The process reduces the cross-section area of the workpiece. Since the volume of the rolled material remains constant, the reduction in cross-section area results in equivalent increase in length. A series of reductions is generally made to produce the workpiece of the desired geometry. The workpiece will generally assume the shape of the roll geometry, which may range from cylindrical to a very complex shape depending upon the desired end product [3,4].

The rail rolling mill located at the Steelton, Pa; plant of Bethlehem Steel Corporation produces a wide range of standard railroad tee rails, girder rails for railway pavement, conductor rails, crane rails and special rail sections. The profile of a standard tee rail is as shown in Figure 1. At the Steelton plant, molten steel produced in electric arc furnaces is first processed through a continuous caster which forms it into 14.6" x 23.6" rectangular cast blooms. These blooms are hot rolled through a two-high reversing blooming mill to produce blooms of smaller cross-section, about 8.5" x 10", prior to transporting them to the rail mill.

The rail mill consists of three mill stands: (1) 35" Mill - Roughing Stand, (2) 28" Mill - 2nd Roughing Stand, and (3) 28" Mill - Finishing Stand. The 35" Mill is a two-high (two rolls) mill, while the 28" Mill is three-high

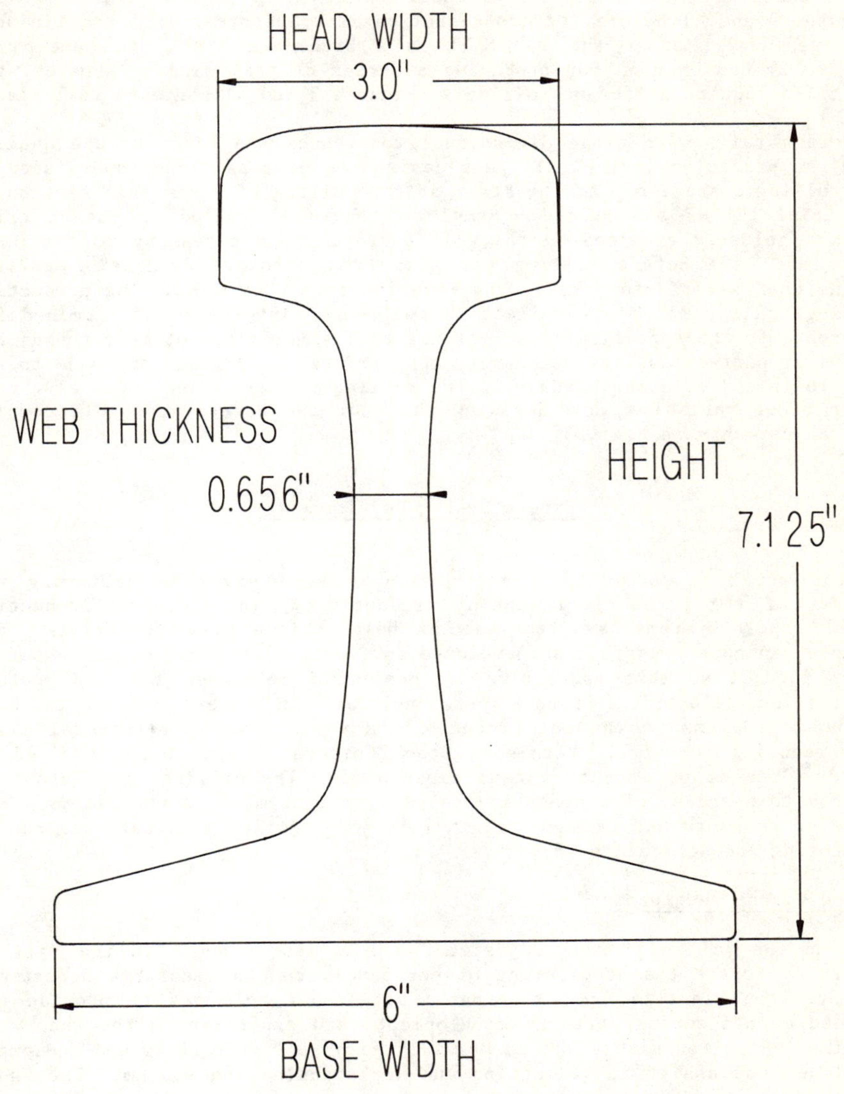

FIGURE 1.   STANDARD TEE RAIL GEOMETRY
(RAIL SIZE: 132RE)

(three rolls). Initially, the rectangular bloom cross-section is progressively developed into the rough contour and shape of the rail section by
rolling through the five grooves (passes) of the 35" Mill Roughing Stand.
The geometries of these five grooves are shown in Figure 2. The workpiece
is further deformed through five additional roll passes at the 28" Mills
(Figures 3 and 4) before its cross-section is transformed into the finished
rail geometry. Since the 35" Mill is a breakdown mill, the same rolls
(Figure 2) can be used for producing a series of rail sizes. However, the
28" mills require different roll sets (Figures 3 and 4) for each rail size.

Producing rails with close dimensional tolerances and high surface quality
requires precision in the roll pass designs as well as close supervision of
the rolling operation. At the start of the rolling of a new rail section, a
few trial blooms are rolled. Sample of the rail rolled from each trial
bloom is closely examined by the mill operators and necessary roll adjustments are made before rolling the next trial bloom. Production rolling
begins when satisfactory rail cross-section is achieved. During production
rolling, rail samples are collected at frequent intervals and examined for
adherence to the specifications. If the rail samples do not meet the dimensional or surface quality standards, appropriate adjustments are made to the
mill rolls and/or other hardware. The rolling mill personnel must also consider other variables when deciding which of the several possible complex
roll adjustments is best [5].

Expert System Development

In the steel industry, much attention has been given to improving the
quality of the product, increasing productivity, and reducing production
cost. Expert systems have been used to help achieve these objectives. For
example, an expert system was developed to forecast abnormal conditions in a
blast furnace so that corrective actions could be taken [6]. In another
application, electrical repair personnel use an expert system to help
diagnose problems in the motor controllers which power a structural mill,
thus reducing downtime. Bethlehem Steel Corporation has been involved in
the development of expert systems since 1985. The original projects were
limited to stand-alone diagnostic systems for equipment in the plants. The
scope of the work has expanded to include more complex applications such as
monitoring and scheduling [5].

<u>Chronology of Development</u>

Work on the rail mill roll adjustment expert system began in the Fall of
1987. A project team consisting of personnel from the Research Department
and Steelton Plant was put together. Research's role was to provide the
knowledge engineering, program development, and expertise in the theory of
rolling. Steelton's role was to provide the practical rolling and equipment
knowledge and assist in validating and implementing the system. The knowledge acquisition process consisted of interviews between the expert operators and Research personnel. A prototype system was developed and reviewed
several times with the mill operators at Steelton for suggestions. The
first useable system was installed in the mill in the Spring of 1988.
However, the refinements and additions to the system continued into the
Fall. The key to the success of the project was the early involvement of
the users, which helped ensure their acceptance and enthusiasm for the
system.

88

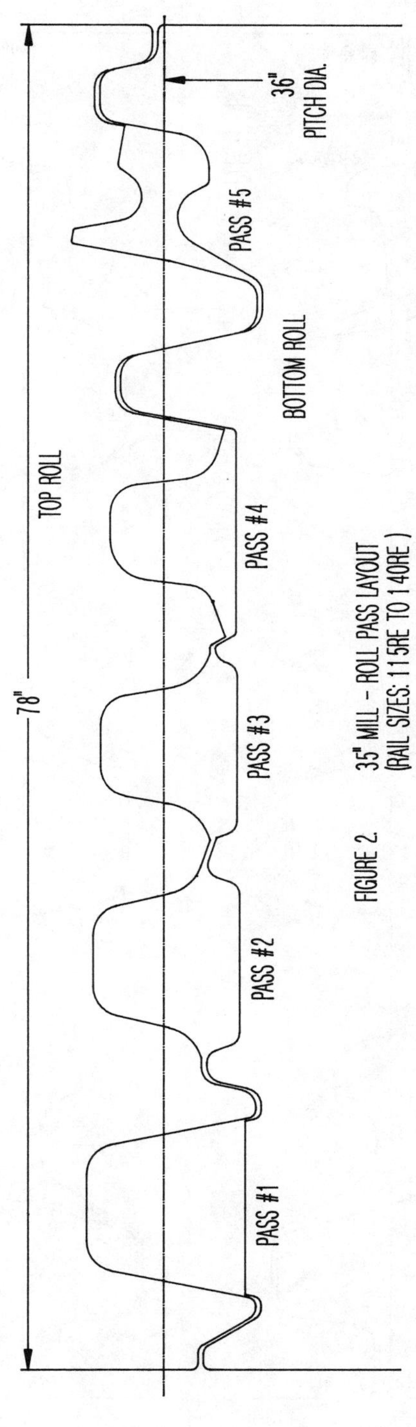

FIGURE 2.   35" MILL – ROLL PASS LAYOUT
(RAIL SIZES: 115RE TO 140RE)

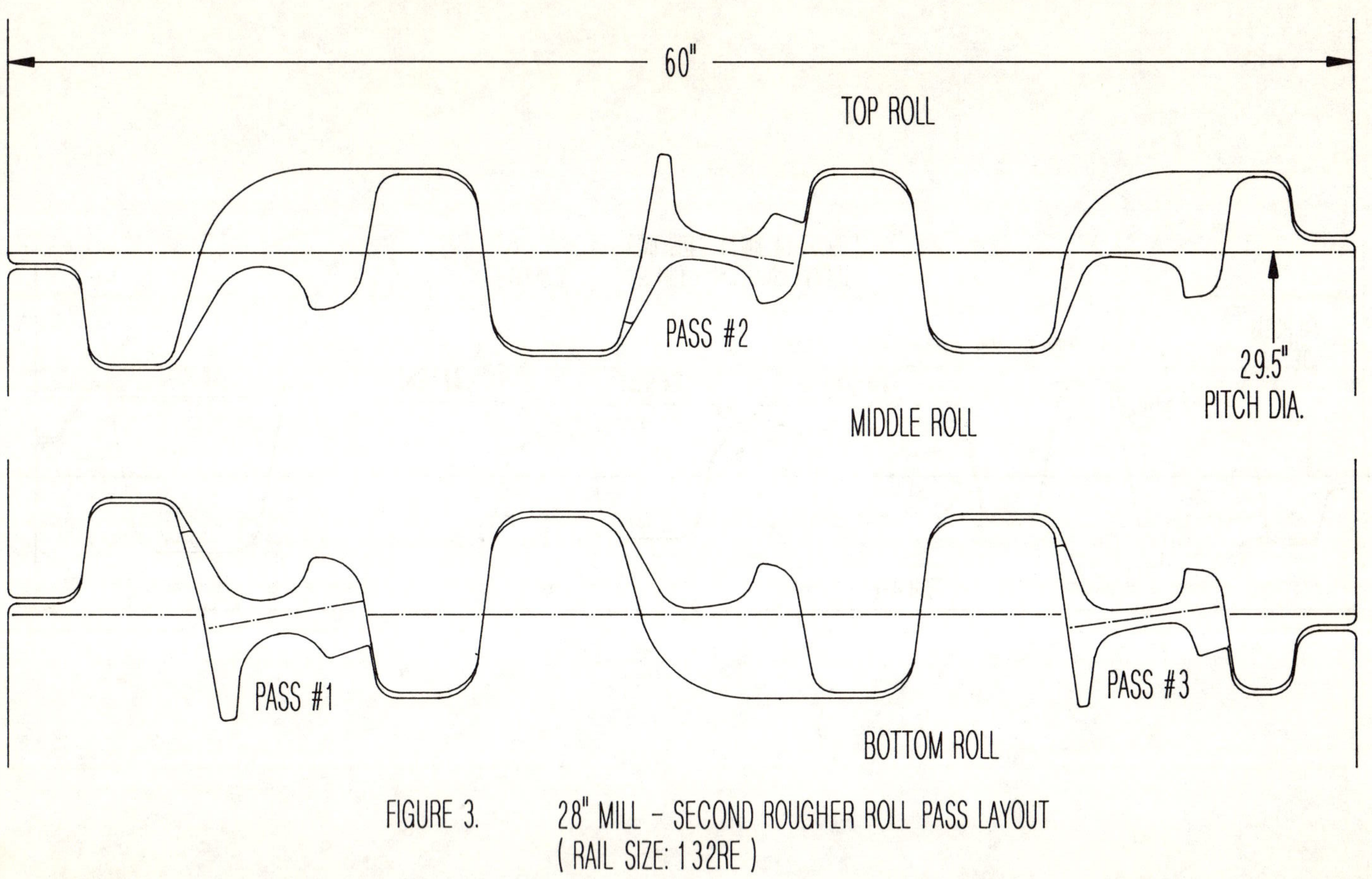

FIGURE 3.   28" MILL – SECOND ROUGHER ROLL PASS LAYOUT
(RAIL SIZE: 132RE)

90

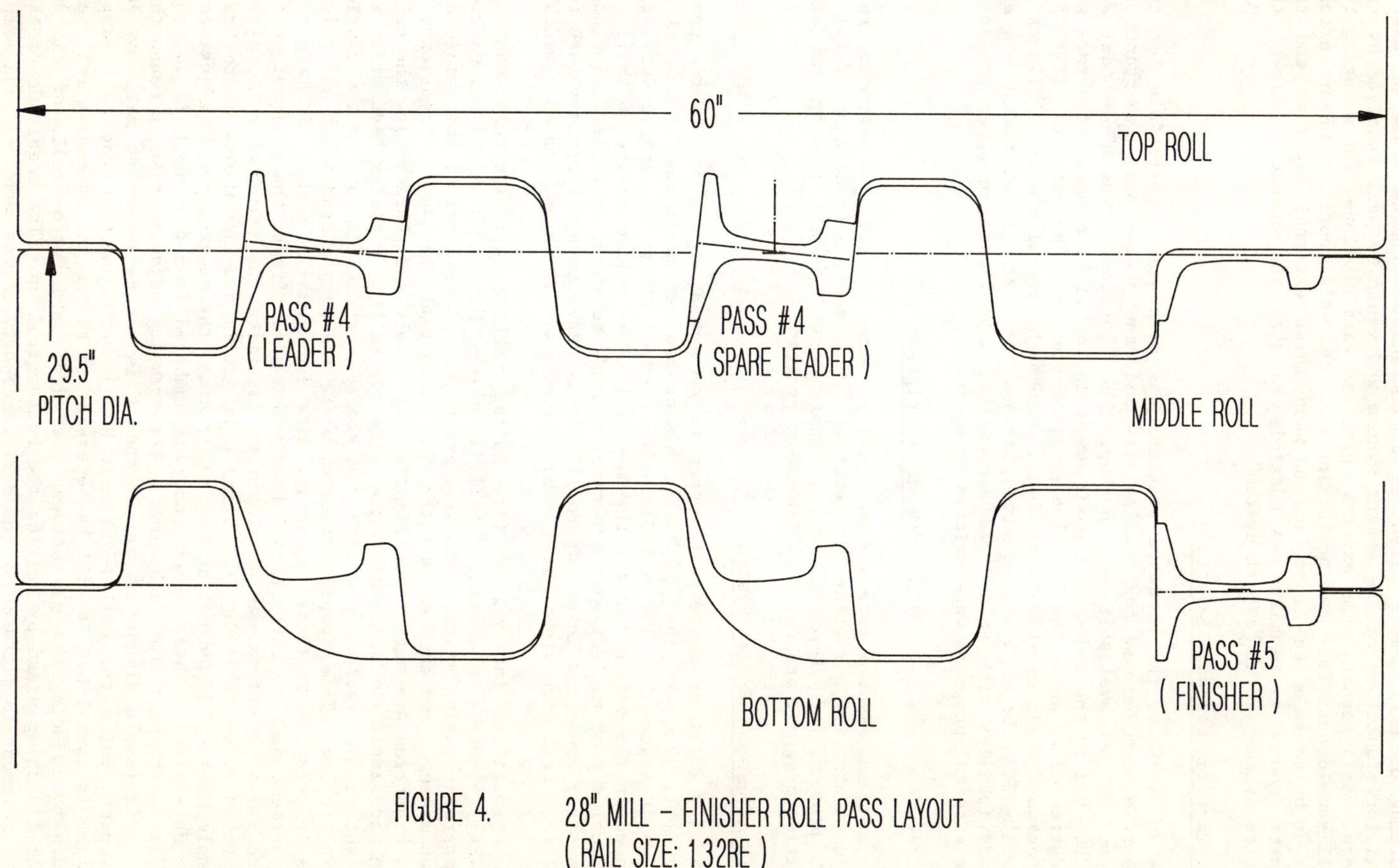

FIGURE 4. 28" MILL – FINISHER ROLL PASS LAYOUT (RAIL SIZE: 132RE)

91

During the early stages of development, it was determined that automatic plotting of rail sample dimension trends and SPC charts would be useful auxiliary functions of the expert system and would be well received by the users. As a result, some external programs were developed for managing the rail sample data base and performing data analysis routines. Every effort was made to make the interface between these external programs and the expert system as seamless as possible so that the operators viewed the entire system as the "expert system".

<u>Software and Hardware Selection</u>

Use of a PC as the hardware platform was determined appropriate for the expert system because the availability and ease of use were important. A second PC was available as a backup. The PC selected was AT&T Model No. 6300. Due to the industrial environment in the mill, a small structure was constructed to house the PC. The expert system shell selected was LEVEL 5 - a backward chaining rule-based shell. Although several similar shells exist for the PCs, Level 5 was selected because the Research Department was already familiar with it and had successfully applied it in other projects. The external programs were written in BASIC.

System Description

The complete system consists of: (1) an expert system to recommend roll adjustments, (2) a routine for plotting rail dimension trend charts, (3) an SPC program for as-rolled rail dimensions, (4) a data listing and data editing program, and (5) a file back-up program.

<u>Roll Adjustment Expert System</u>

The flow chart of the expert system to recommend adjustments to the upper and lower rolls of the 28" Mill finisher and second rougher stands is as shown in Figure 5. The rail dimension control limits (tolerance data) are stored in the system for all the tee rail sizes rolled on the 28" Mill. A typical set of rail dimension control limits, as shown in Table I, includes aim size, upper and lower control limits and maximum and minimum values for all four rail dimensions (head width, height, web thickness and base width).

At a specified interval of time during rolling (approximately every 20 rails), a sample of as-rolled rail is taken. The sample is cooled to room temperature using standard cooling practices. Dimensions of the sample are then measured and entered into the PC. Along with the sample dimensions, 16 other entries are made which identify the sample and characterize the rolling and sample cooling conditions. The system then compares head width and height of the rail sample with the stored control limits for the section being rolled. This comparison shows where the sample head width and height are in relation to their respective aims and control limits. Since a roll adjustment made to correct rail head width also affects the rail height and vice versa, a matrix was developed to determine the required roll adjustments as shown in Table II. Using this matrix and deviations of the rail sample head width and height from their aims, the necessary roll adjustments are determined. These adjustments, as numbered from 0 through 8A in Table II, are different from each other. For example, adjustment No. 0 means that the rail sample dimensions are OK and it is not necessary to make any adjustments while the roll adjustment No. 1 indicates that the rail sample height is normal but it will be necessary to reduce the excessive rail head width by reducing the gap between the middle and bottom rolls of the 28" Mill finishing stand by pulling the bottom roll up. The magnitude of this roll movement is dictated by the difference between the measured rail sample

92

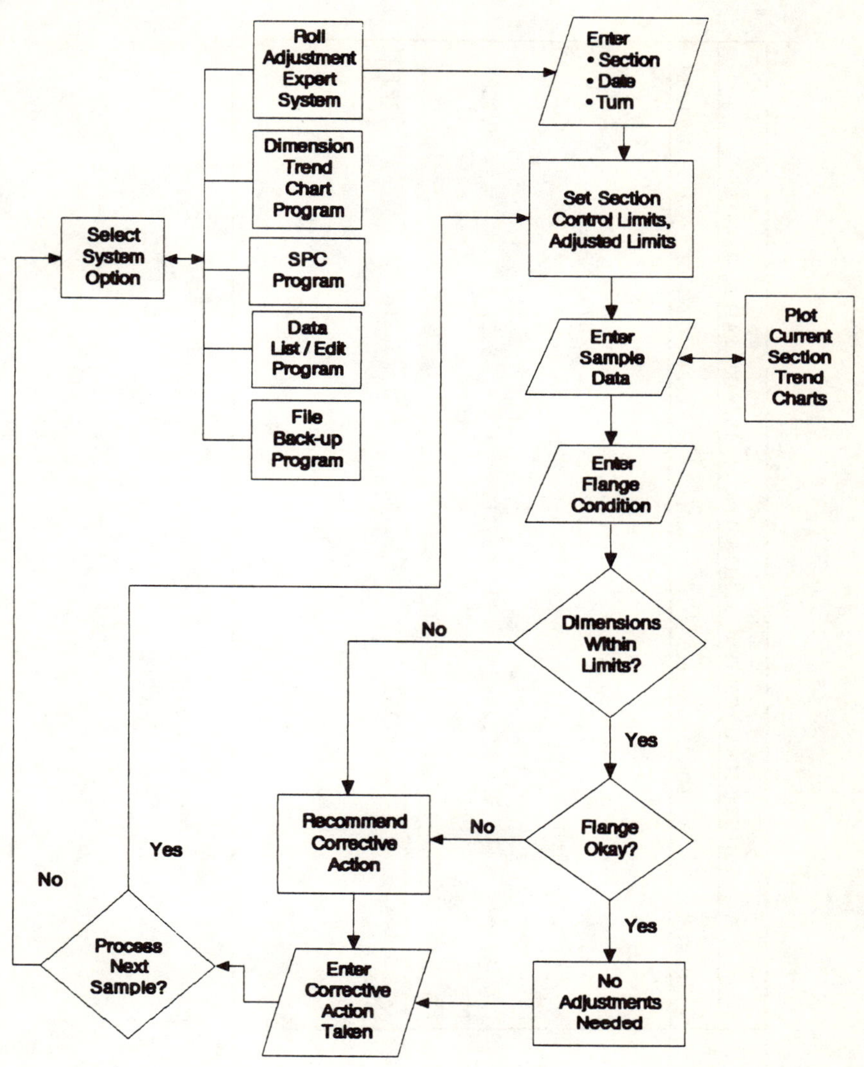

FIGURE 5. SYSTEM FLOWCHART

TABLE I. - TYPICAL SET OF RAIL DIMENSION CONTROL LIMITS

RAIL DIMENSION	DIMENSIONAL CONTROL LIMITS FOR 85 lb/YARD RAILS, INCHES				
	MINIMUM	LOWER CONTROL	AIM	UPPER CONTROL	MAXIMUM
HEAD WIDTH	2.548	2.553	2.563	2.573	2.578
HEIGHT	5.203	5.208	5.213	5.223	5.228
BASE WIDTH	5.148	5.153	5.188	5.218	5.223
WEB THICKNESS	0.548	0.553	0.563	0.593	0.598

TABLE II. - 28" MILL ROLL ADJUSTMENT LOGIC FOR RAIL HEAD WIDTH & HEIGHT

	Head Width Narrow	Head Width Approaching Lower Control Limit	Head Width O.K.	Head Width Approaching Upper Control Limit	Head Width Wide
Height – Excessive	7	7A	3	3A	5
Height Approaching Upper Control Limit	7B	0	0	0	5A
Height – O.K.	2	0	0	0	1*
Height Approaching Lower Control Limit	8A	0	0	0	6A
Height – Short	8	4A	4	6B	6

Left axis (HEIGHT, top to bottom): Upper Control Limit, Near Upper Control Limit, Near Lower Control Limit, Lower Control Limit

Bottom axis (HEAD WIDTH, left to right): Lower Control Limit, Near Lower Control Limit, Near Upper Control Limit, Upper Control Limit

* Numbered recommended roll adjustments (for explanation refer to System Description)

head width and the aim head width. The suggested roll adjustments, presented on the computer screen, include both the direction and magnitude of roll
movements.

If there are no corrective actions required for head width and height, the
system evaluates the need for any corrective actions for the top and bottom
flange lengths as shown in the matrix in Table III. These adjustments are
numbered from 9 through 16 and are different from each other. The flange
length corrective actions, if needed, are then presented on the computer
screen. Since web thickness is expected to be acceptable when the other
dimensions are within tolerances, no corrective actions are recommended for
the web thickness. The operator is then given an opportunity to enter the
roll adjustments made, if any, to the 28" Mill finisher and/or second
rougher stand rolls. The system is then ready to process the next sample.

Rail Dimension Trend Charts

The rail dimension trend charts can be accessed in two ways: (1) from the
initial entry into the system and (2) from the roll adjustment expert system
as shown in the flow chart in Figure 5. If the first option of reviewing
the trend charts is chosen, the operator will then be prompted for the
section, date of rolling and shift I.D. This option allows review of any of
the data stored in the system. Accessing the trend chart plotting routine
while in the roll adjustment expert system (option 2) allows review of the
trend charts of the current shift as well as those of the past shifts on the
section being rolled. This option does not allow review of trend charts of
any section other than the one being rolled at the time. Separate trend
charts are plotted for each of the four rail dimensions.

SPC Program

The SPC program provides SPC charts for each of the four sample dimensions
(i.e. head width, height, base width, web). As in the case of dimensional
trend charts, the measurements included in the SPC charts are only from the
first measurements of water box cooled samples. Measurements of the air-
cooled and back-end samples as well as repeat measurements are not used in
the SPC calculations. After the user selects a section, rolling date,
starting turn and dimension, an SPC chart will appear on the screen. The
SPC chart consists of:

- a table of sample dimensions and SPC calculations, and
- plots of the mean and range along with horizontal lines indicating SPC
  control limits for that section.

Data Listing/Editing Program

The Data Listing and Editing program provides two reports for viewing and
printing sample information. The first report lists all the information for
a sample, including the roll adjustments recommended and actions actually
taken. An editing option is provided from this report which allows the user
to edit sample information. The second report contains a reduced set of
sample information. Both reports show an entire shift of sample information
for a section.

File Back-Up Program

The File Back-up program can be used to back-up section files. There are
two files associated with each section, a data file and a master file.

TABLE III. - 28" MILL ROLL ADJUSTMENT LOGIC

FOR RAIL TOP FLANGE & BOTTOM FLANGE

<table>
<tr><td rowspan="7">T
O
P

F
L
A
N
G
E</td><td>Top
Flange
Long</td><td>15</td><td>14</td><td>16</td></tr>
<tr><td>Top
Flange
O.K.</td><td>9 *</td><td>0</td><td>10</td></tr>
<tr><td>Top
Flange
Short</td><td>12</td><td>11</td><td>13</td></tr>
<tr><td></td><td>Bottom
Flange
Short</td><td>Bottom
Flange
O.K.</td><td>Bottom
Flange
Long</td></tr>
</table>

BOTTOM FLANGE

* Numbered recommended roll adjustments

These files can then be transferred to other applications for further data analysis. The user specifies the section to be backed up after which both files are copied.

Implementation

During implementation of the system, the operating environment and plant culture created the usual "resistance to change", which is encountered when new technology is introduced into the workplace. The users are primarily senior employees with very little or no experience with computers or electronics in the workplace. Typical working tools utilized are hammers and wrenches for making equipment adjustments and micrometers, calipers, pencil and paper for measuring and recording rail section sample data.

The rail mill and plant had been in a major quality improvement mode for at least two years before the system was implemented. During this sustained quality improvement period, improved equipment, SPC charts, and standard operating procedures had been instituted. These changes helped set the stage for another quality improvement project such as the roll adjustment system. Add-on auxiliary programs to automate data plotting and charting, which were previously done manually, helped give the operators an immediate tangible benefit from accepting this system.

About 10 to 12 operators were trained in the operation of the roll adjusting system. Training was done one-on-one away from the job site with the assistance of an enthusiastic experienced operator. This operator was a member of the rail mill quality improvement team and had the respect of other operators because of his overall mill knowledge and 35 years of experience. Additionally, a detailed operating procedure was developed and given to each operator as a guide to running the system. The operating procedure was complete with copies of the system data input and graphics screens. When the system was implemented on the shop floor, a knowledgeable system operator worked with the operators on each shift for a week to facilitate implementation and to keep track of the suggested improvements made by the operators. A list of 10-15 suggestions to make the system easier or more comfortable to use was developed. These suggestions ranged from changing a word on a screen heading to modifying the software to make it easier to switch back and forth between data input and graphics screens. No matter how trivial or unimportant the suggestion made by the user appears to be, if that idea is taken seriously and acted upon, a sense of system ownership is developed which makes acceptance that much quicker and easier.

Summary

The overall strategy for improving product quality on the rail mill has been focused on equipment design and improvement, operator job training, and improved process control and understanding. As part of these efforts, a PC-based expert system has been implemented to recommend roll adjustments during rolling of rails on the 28" Rail Mill and to compile as-rolled rail sample dimension statistics. The system is helping to standardize the mill roll adjustment procedures and to improve the quality of operator decisions. The automatic plotting of rail dimension trend charts and SPC charts compared to manual plotting saves time, provides greater assurance that rails meet the required specifications and results in productivity and yield improvements. The mill operators now enjoy using this system. It has also enriched their jobs, which is a significant factor. The system is being enhanced to add rules dealing with the interaction of other process variables affecting the way mill rolls are adjusted to further improve rail dimensional uniformity.

Acknowledgements

The authors wish to thank all of their associates at Bethlehem Steel Corporation who participated in development and implementation of this roll adjustment expert system. Special thanks are due to the experienced mill rollers J. Gasparovic and D. Goodling for their contributions in developing the roll adjustment rules, A. H. Irwin, Jr., R. C. Sprow, and G. Murphy for their support and guidance throughout this project and F. Verbos for his leadership in carrying out the operator training. The authors appreciate the enthusiastic acceptance by all the system users who made the implementation possible. The authors are also grateful to B. J. Votral for her assistance in preparing the manuscript.

References

1. R. J. Brachman, "The Basics of Knowledge Representation and Reasoning", AT&T Technical Journal, 67 (1) (1988).

2. D. E. O'Leary, "Methods of Validating Expert Systems", _Interfaces_, 18 (6) (1988), 72-79.

3. W. L. Roberts, _Hot Rolling of Steel_ (M. Dekker, Inc., 1983).

4. E. E. Brayshaw, _Rolls and Rolling_ (Blaw-Knox Company, 1958).

5. Bethlehem Steel Corporation, Internal Reports, File Memos, and Informal Communications by M. M. Vyas, D. J. Renn, S. M. Harpster, and A. H. Irwin, Jr.

6. Y. Tsunozaki et al, "An Expert System for Blast Furnace Control at Fukuyama Works", _Nippon Kokan Technical Report_, Overseas No. 51 (1987), 1-10.

WELDEXCELL: A Blackboard System for Off-Line Intelligent Planning and Control of Robotic Welding

J. E. Jones
American Welding Institute
10628 Dutchtown Road
Knoxville, TN 37932

Colo. School of Mines
Center for Artificial Intelligence
Golden, CO 80401

D. R. White
MTS Systems, Incorporated
Advanced Technology Development
14000 Technology Drive
Eden Prairie, MN 55344

ABSTRACT

Construction of any large structure can require many thousands of feet of welding. Whenever the welding process can be streamlined or automated, tremendous cost savings can be obtained. The WELDEXCELL system is a <u>WELD</u>ing <u>EX</u>pert manufacturing <u>CELL</u> that provides computerized technical support information, off-line weld planning, and an integrated welding robot/welding system/vision system controller. The first of two sub-systems, the Welding Job Planner (WJP) accomplishes off-line intelligent weld planning for both automated and manual welding processes. The second subsystem, the Welding Job Controller (WJC) provides a fully integrated hardware control environment with associated software for combined control of a welding robot, welding equipment, and a robotic vision system. In the WELDEXCELL system, a series of expert systems and databases have been combined in a new type of computer software environment called a blackboard. There are as many as 19 separate components of the Welding Job Planner subsystem of WELDEXCELL which fall into five interrelated functional groups. WELDEXCELL will be used by design engineers, welding engineers, mechanical engineers, and NDT engineers for both manual welding and to interface to automated and robotic welding systems and vision systems. WELDEXCELL also includes the control system hardware and software to provide off-line intelligent adaptive control of the welding process itself.

The development of WELDEXCELL is a multi-year effort involving a partnership of government, industry, university research, and technology transfer. The project has already generated new concepts with potential for future spin-off benefits. The ultimate payback in productivity will be large for the American welding, fabrication, manufacturing, and construction industry in general.

Expert System Applications in
Materials Processing and Manufacturing
Edited by M.Y. Demeri
The Minerals, Metals & Materials Society, 1989

OVERVIEW

The American Welding Institute (AWI), together with the other WELDEXCELL team members, the Colorado School of Mines (CSM), and MTS Systems, Incorporated (MTS), is developing an intelligent weld process planner for flexible welded fabrication known as the <u>WELD</u>ing <u>EX</u>pert manufacturing <u>CELL</u> (WELDEXCELL). This project entails the development of a computerized blackboard with a series of linked expert systems acting as a welding engineers' assistant, and software to download welding procedures from the weld designer to a welding workcell for automatic execution of the planned welds. The system will also employ sensors to record actual weld process parameters and a postweld analysis capability to examine these parameters and update the welding procedure between passes. These sensors include a seam tracker which will provide path corrections to the welding robot during a weld.

Many parts of the system software have already been developed, and some of the software is commercially available as individual expert systems and databases. But, the heart of WELDEXCELL is a new computer architecture called a "blackboard". This blackboard system allows the interconnection of multiple expert systems and databases with a central goal. WELDEXCELL will be one of the first commercial computerized blackboard applications ever developed.

BACKGROUND

The joining of metals into fabricated components and structures is a difficult task. The most common method of joining metals is welding, but the welding process is complex and requires several important steps to be performed in a carefully integrated manner. Although the process may seem simple to an experienced welding engineer, when analyzed in sufficient detail, the engineering/planning processes are extremely complex. Such an analysis was performed in developing the task description for the WELDEXCELL Welding Job Planner (WJP). The weld joint is first designed and engineered properly, then that design must be correctly communicated to the fabrication facility. The appropriate welding consumables, including filler metal and protective flux or inert gas, are chosen. Then the welding procedure is specified, including preheating schedules; welding variables such as voltage, current and travel speed; and postweld heat treating. Finally, the weld must be performed under highly skilled human guidance and control. A minor error in any of these steps, if undetected, can create an unsuitable welded component, which in later use may result in a catastrophic failure and perhaps loss of life.

An extremely complex and interrelated system of codes, specifications, tests, and inspections ensures that the vast majority of welds will never fail in service. Fortunately, a large number of engineers, designers, and welders work within the system of codes and specifications to ensure the high quality of welded joints, but this system is very expensive and requires the careful attention of many human experts. Consequently, welding is an ideal application for computerized expert system technology. However, no single expert system could be expected to perform the myriad of tasks required to make a welded joint. For example, there are over 100 welding processes ranging from simple flame heating to exotic laser welding; there are several hundred welding filler metals -- from plane carbon steel to elaborate chemical mixtures of alloying ingredients; and there are over 1000 different grades of weldable steels classified by the American Society for Testing and Materials (ASTM). The possible combinations of welding process, filler metal, and steel base metal would number into the millions.

The expert systems needed for welding include materials selection, joint design, welding process and procedure selection, and a CAD system interface to draw the design and communicate that design to the welder. The ultimate goal also includes an intelligent system to instruct a complex welding workcell to perform the weld; and a workcell simulator to allow off-line automated weld planning.

WELDEXCELL SYSTEM BLACKBOARD

It is clear that the type of distributed problem-solving in multiple knowledge domains involved in this multidisciplinary engineering problem cannot be addressed using a single knowledge source (KS). Rather, multiple knowledge sources and humans will cooperate to solve a broad problem. The technique to be applied to this data and knowledge-integration problem is the computer blackboard architecture.

The concept of blackboard architectures was discussed in the literature a early as 1962; however, no applications were built until the 1970s. A blackboard is being used for this expert integration environment because it possesses capabilities to support problem solving while accounting for diverse types of information, methods for combining various types of data while resolving conflicts, and the ability to accommodate different program modules without requiring a complex interface.

The problem solving technique which has been applied to the blackboard model is to divide the problem into loosely coupled subtasks which are then operated on by specialized programs with access to various information sources. The information sources consist of knowledge bases, expert systems, databases, and interfaces to human experts. The advantage of such a system is that much larger quantities of information can be used in a fully integrated manner to solve the problem and develop the weld plan. The human experts supply the external information about the required welding task and then review the intermediate and final plans. The system also includes facilities to query a human expert in the event that conflicts outside of the system's domain of expertise occur. The time required by a human expert will be substantially reduced, thus allowing more design and planning to be accomplished with higher overall quality and reliability by the same number of human experts. Also, the system will reduce the time required to test quality and practice automated welds. This will substantially reduce the problem of small batch size automation.

The blackboard software architecture is analogous to a group of experts seated before a blackboard, with only one expert allowed to approach the blackboard at a time. A monitor is empowered to call on the experts individually to modify the blackboard's contents. Following each contribution, the monitor evaluates the state of the blackboard's contents and, based on its planning algorithms, considers which expert to call on next. If the "experts" described in this scenario are replaced by knowledge sources (KS's), which include expert systems, databases, knowledge bases, human users, graphical data and information, etc., a blackboard system results. The monitoring and control functions are performed by what is essentially another expert system with planning algorithms designed to move the expert system toward a problem solution.

The blackboard's purpose is to provide a framework for the interaction of the multiple independent knowledge sources and to respond opportunistically to the changing contents of the blackboard to achieve a solution. There are eight behavioral goals for the intelligent blackboard control system to accomplish this task. They are as follows:

1. Make explicit control decisions that solve the control problem of multiple independent knowledge sources.

2. Decide what actions to perform by determining what actions are desirable and what actions are feasible.

3. Adopt variable task size control heuristics.

4. Adopt control heuristics that focus on action attributes which are useful in the current problem solving situation.

5. Adopt, retain, and discard individual control heuristics in response to dynamic problem solving situations.

6. Decide how to integrate multiple control heuristics of varying importance.

7. Dynamically plan strategic sequences of actions.

8. Reason about the relative priorities of domain and control actions.

The blackboard controller controls the blackboard, monitoring the activities of the knowledge sources attempting to find a solution to the weld design problem. At various levels ranging from abstract to very detailed, decisions are made such as which problem to solve next, whether forward or backward chaining reasoning is to be used and which knowledge source to activate. While building a master expert system to control the problem solving blackboard is a complex solution, it provides the flexibility to solve both broad planning problems and perform detailed scheduling.

The blackboard control system contains more explicit support for meta-level facilities. The blackboard is divided into multiple partitions which contain classes. The classes contain objects. The objects, which contain the data used by the KS's to solve a problem are placed in the blackboard by KS's or by external processes such as human interactions or interaction with the databases.

Another concept for organizing problem solving with multiple, diverse cooperating sources of knowledge is being applied to the blackboard. A hypothesize-and-test paradigm is a mechanism which can provide a high degree of cooperation among the knowledge sources. Thus, the solution finding is an iterative process, which involves two steps:

- Create a hypothesis (an educated guess about some aspect of the   problem)

- Test the plausibility of the hypothesis

As the blackboard proceeds toward a solution, the system will build on the knowledge about the problem contained in its knowledge sources and the changes in the state of the system knowledge (i.e., in the contents of the blackboard) produced by previous hypothesis. This iterative process ends when the contents of the blackboard form a consistent hypothesis which satisfies the requirements of an overall weld design solution.

WELDEXCELL SYSTEM DESCRIPTION

WELDEXCELL is logically divided into two major subsystems: the Welding Job Planner (WJP) and the Welding Job Controller (WJC). A high level block diagram is shown in Figure 1.

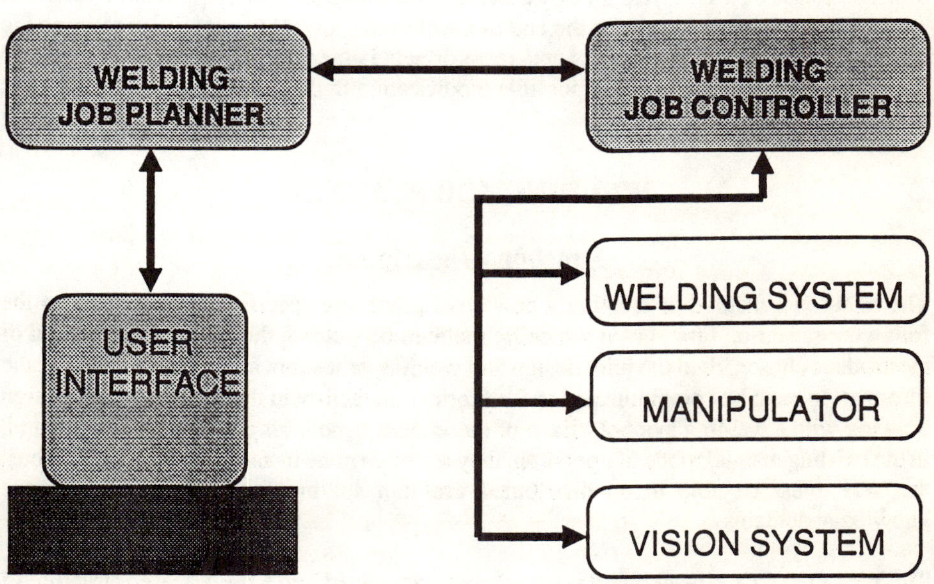

Figure 1. High level block diagram of the WELDEXCELL System

<u>**Welding Job Planner**</u>

The WJP will use various expert systems, knowledge/databases, and user input to solve welding engineering design problems. The user will interact with the computerized knowledge resources and the computerized "blackboard" to design the joint and then locate or assist in developing an appropriate welding path and procedure.

The WJP configures this information in the form of a Job Description (weld schedule) which is in turn passed to the WJC for the actual execution of the weld. The user interface for the WJP is primarily the blackboard output file, which is represented as a Welding Procedure Specification (WPS) and displayed on the screen. As information is determined by the expert systems and other systems, the onscreen WPS will be updated. The display shows initial information (such as material type and joint geometry) and the evolution of the WPS as it is developed, including the joint design, robot path planning, and simulation information.

<u>**Welding Job Controller**</u>

The WJC is responsible for ensuring that the various equipment used for the weld, including the welding power supply, the manipulator, the vision system, and other support equipment are coordinated as appropriate to execute the weld schedule from the WJP. During the course of a weld, the WJC will record several weld process parameters (such as voltage, current, wire feed rate, gas flow, travel speed, and temperature as appropriate to the type of welding taking place) as well as any offsets between the planned path and the actual seam location. This data is saved for later inter-pass analysis by the WJP, but the seam offsets will also be used in real time by the controller for adjustments to the robot's planned trajectory (i.e seamtracking). At the end of a weld pass, a weld results file containing the recorded weld process parameters and seam offsets is prepared by the WJC and passed back to the WJP for analysis and possible modification to the job description for the next pass.

WELDING JOB PLANNER

<u>Functional Description</u>

The traditional method of creating a new weld procedure specification is similar to the following scenario. First, given a specific metal to be welded, the welding filler metal or electrode is chosen, then the joint design and welding procedure are selected. Finally, the information must be communicated to the fabrication facility in the form of a joint design drawing with a welding symbol. Each of these tasks is not completely independent and, in the existing manual mode of operation, they are often done in an iterative manner. Thus, the WJP must be able to do distributed problem solving in multiple simultaneous knowledge domains.

The WELDEXCELL blackboard is functionally organized into a frame-based structure of sub-blackboards. Each of the sub-blackboards has a specific functional use in the overall system. These sub-blackboards are described by a series of attribute (i.e. specific goal

Welding Procedure Database (WPSDB)

The welding procedure database consists of process, material, and parameter information for the weld. This database is structured in accordance with the proposed AWS/ANSI Standard A6.1-90 for welding procedure specifications.

Procedure Qualification Record Database (PQRDB)

This database consists of information similar to WPSDB, but includes test results. Full PQRs are stored in this database conforming to the proposed AWS/ANSI Standard A6.1-90.

Electrode Database (ELDB)

Contains data about welding electrodes and filler metals. The information includes not only the designated AWS A.5 standards but also manufacturer published data. The records include typical/recommended usage, composition, operating parameter ranges, etc.

Steels Database (MATDB)

Contains ASTM, AISI, ACI, and UNS weldable steel information, including composition and mechanical properties. Recommended electrode/filler metal usage data is also incorporated.

Heat Treatment Database (HEATDB)

Contains pre- and postweld heating schedule information. Includes data for carbon equivalent (CE), P_{cm}, and PH_a analysis, as well as WRC-published recommended temperatures/times.

Joints Database (JOINTDB)

Contains welding joint design detail information, including data for root opening, included angle(s), tolerances, etc. Currently, the data is not CAD compatible; but if necessary for system operation, the data will be converted.

Welding Symbols Database (SYMDB)

Contains AWS/ANSI D2.4-86 standard welding symbol information for developing standard welding symbols on mechanical drawings. This data is already CAD compatible and will be maintained in Navy CALS compatible format.

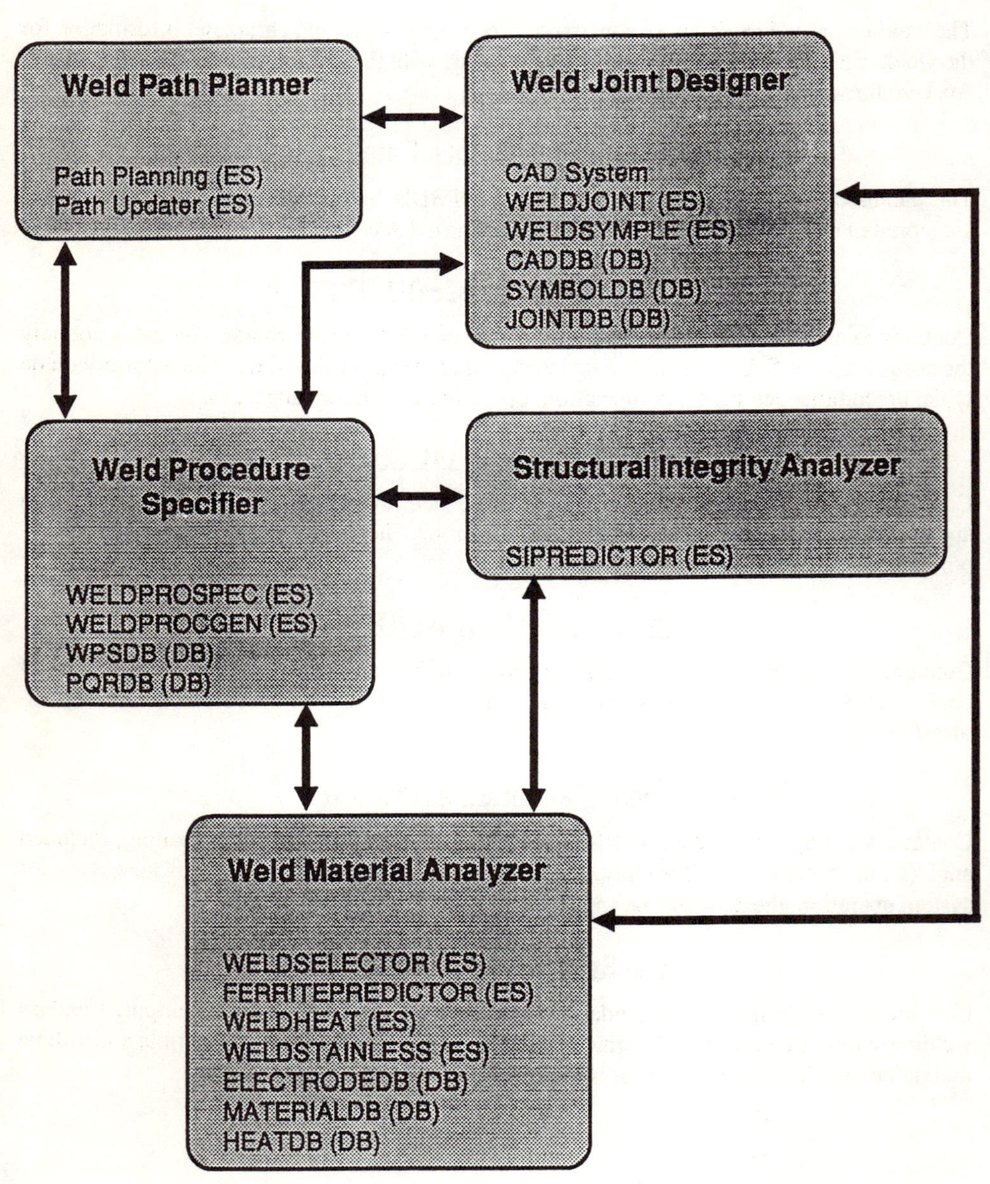

Figure 2. Overall Breakdown of the Welding Job Planner Subsystem

parameters) values which are determined, through operation of the system, to the greatest extent possible whenever that sub-blackboard is instantiated during a consultation. Each sub-blackboard also inherits the attribute values of the parent (main) blackboard. The blackboard structure has the ability to go through a scenario repeatedly with minor changes so that engineering iterative solutions ("what-iffing") techniques can be applied.

The major components (sub-blackboards and other routines) of the WJP are the Joint Designer, the Structural Integrity Analyzer, the Material Analyzer, the Procedure Specifier, the Path Planner and the Main Blackboard. Each of the sub-blackboards are made up of several cooperating expert systems and databases. Each of these major components is described below in greater detail.

Most of the expert systems and databases have already been developed or prototyped by AWI and the Colorado School of Mines (CSM) as part of the Welding Information Network (W.I.N.^{TM}) system. In some cases AWI is commercially marketing, a wide range of expert systems for use in welding engineering decision support. In development of these expert systems, it was important to use recognized standards, codes, and existing tested procedures in the operation of the expert system. Two principal organizations in the United States are primarily responsible for the technical standards and procedures of welding (besides the United States government): the American Welding Society (AWS), and the Welding Research Council (WRC). Selected committees of each organization were approached to enlist their cooperation and input with respect to the development of the expert systems. These committees have supplied expert knowledge, evaluation, and beta test sites for the prototyped expert systems. Future beta testing will also be performed by U. S. Navy shipyards. In addition to the expert systems, several technical databases are used by the Welding Job Planner. Each database will be structured as summarized in Table I. Figure 2 illustrates the overall breakdown of the WJP.

Weld Joint Designer

The Joint Designer Sub-blackboard (JDS) functions to prepare a welding joint design and communicate the joint design information in a graphical format to the engineer, as well as to the shop floor. Standard joint design formats (consistent with the AWS standard graphical description) are being employed for the basic joint design. Also, the welding symbol prepared by JDS conforms to AWS/ANSI standard A2.4-86.

WELDSYMPLE

A large amount of information is needed to describe a weld procedure on a mechanical drawing. The welding technique and testing must be specified as well as the joint design and machining requirements. A shorthand way of describing a weld, known as a welding symbol, is used on a mechanical drawing to describe the specified weld. The technique for developing a weld symbol is like that of constructing a word in the English language. A set of symbol elements (the alphabet for the word construction analogy) is available. By choosing appropriate symbol elements and assembling them in an appropriate manner, a symbol (a word in the English analogy) can be constructed. There is a nearly unlimited number of symbol element combinations which could be used to generate welding symbols, so generating a CAD library of symbols is, for general application, not practical. The appropriate weld symbol must be generated each time it is to be used.

The expert system WELDSYMPLE is designed to use a symbol base (database of graphic welding symbol information) and input from a human user or the blackboard regarding the weld joint design/application to draw the appropriate welding symbol using a CAD system. WELDSYMPLE uses the same logic processes which would be applied by a welding engineer to develop a welding symbol. The symbol is generated according to the rules established by AWS in documentation reflecting the standardized use of welding symbols (AWS A2.4-86 "Standard Symbols for Welding, Brazing, and Nondestructive Examination"). This expert system is currently available as a stand alone system and was tested by AWS prior to its commercial release.

WELDJOINT

The design of a welded joint is, as with many engineering decisions, a delicate balance of compromises. The joint design attempts to combine several criteria simultaneously, some of which may conflict. The joint must be machined to allow sufficient clearance for the welding operation, but with a minimum of open space to fill with the expensive filler metal and time consuming welding operations. The design must accommodate the configuration of the structural shapes to be joined, but also minimize the stresses which occur on the joint in service and the residual stresses that develop as the weld shrinks due to non-uniform temperature distribution during solidification and cooling.

The WELDJOINT expert system interacts with a graphics-based data system to produce a drawing of the weld joint. In addition, the expert system provides much of the output information so that the welding symbol can be produced for the mechanical drawing. The data and graphical layout of the figure are in accordance with AWS/ANSI standard joint design, as described in AWS D1.1, "Structural Welding Code."

Structural Integrity Analyzer

The Structural Integrity Analyzer Subblackboard (SIS) will function to provide the design engineer with basic structural integrity information and help to set the NDE criteria. The system utilizes the basic Linear Elastic Fracture Mechanics (LEFM) criteria for defect size limit setting. However, the SIS will have provisions to include more complex analyses in future implementations.

SI-PREDICTOR

A structural integrity analysis is a necessary part of the overall system. The SI-PREDICTOR expert system provides a fracture mechanics approach to the analysis of structural safety. The system utilizes basic information about the structural geometry type, mode of loading, and structural dimensions, and also about the material being used (tensile properties and fracture toughness). SI-PREDICTOR is currently based upon a LEFM approach; consequently, solutions are checked to see whether the limits of linear elasticity are violated. SI-PREDICTOR determines a critical defect size for a structural component geometry. Six component geometries are available: Vessel, Truss, Plate, Beam, Girder, and Pipe.

The SI-PREDICTOR expert system program calculations have been verified independently for numerous test cases to ensure that the program is error free. The accuracy of the critical defect sizes calculated are dependent only upon the accuracy of the input; component dimensions, applied loading, and material properties.

Material Analyzer

The Material Analyzer Sub-blackboard (MAS) will enable the design engineer to perform optimum selection of welding consumables and to set pre- and postweld heat treatment (PWHT) to optimize the weld properties. This system utilizes all applicable Mil. Spec's. and standards, in addition AWS welding electrode and filler metal specifications are included in the databases. The weld preheat and PWHT will be based on Mil. Spec's. and Welding Research Council (WRC) published guidelines. In addition, the MAS will provide the user with the latest state-of-the-art technology for analysis of special nonstandard materials, or universal weld heating requirements.

WELDSELECTOR

The selection of a welding electrode or filler metal is a complex task requiring detailed information about the base metal to be welded and the properties of the electrode and filler metal. In addition, several aspects of the welding operation must be examined and decisions made regarding the specific application in order to narrow the list of possible electrode choices. A weld, which is a small bit of solidified metal, is expected to have the same (or perhaps better) properties as the base metal that it joins. The base metal may have undergone hours of careful and expensive heat treating and processing, yet the weld metal must be as corrosion resistant, as strong, as ductile, and as fracture resistant as that base metal.

The WELDSELECTOR expert system is able to access data about the welding materials through the use of extensive databases which contain information about base metals and electrodes. The base metals database currently contains over 1,000 grades of steel identified by ASTM classification. Navy-used steels are being added to the database, including the HY and HSLA steels commonly used in shipbuilding. The electrode database contains all of the AWS classified welding electrodes which are used in the United States for three welding processes: Shielded Metal Arc Welding (SMAW), Gas Metal Arc Welding (GMAW) and Flux-Cored Arc Welding (FCAW). The various military qualified electrodes are currently being added to the cross-reference listing.

WELDSELECTOR follows the logic processes which are used by a human expert to determine an appropriate filler metal. Given basic information regarding the material to be welded and using the databases of the base metals and electrode properties, an initial feasible list of electrodes is produced. The list is then ranked based on decision factors about the required weld. Examples of the type of decision factors include: (1) the type of welding equipment to be used, AC or DC (to partially determine the chemical design of the welding flux); (2) the degree of hydrogen contamination coupled with the sensitivity of the base metal to hydrogen damage; and (3) the position in which the weld is to be made (e.g., flat, vertical, or overhead).

WELDSELECTOR only uses decision factor data as necessary. The system can prompt the human user for more details when needed. As with any decision making process, conflicting input must be weighed and evaluated based on its resulting impact. This is accomplished in WELDSELECTOR with the use of a numerical rating system of certainty factors (CFs). WELDSELECTOR produces a CF-ranked list of electrodes from which the top choices are selected. These top choices can then be used in the design of an overall welding procedure.

WELDHEAT

The arc welding process often requires additional heat treatment in the form of applying external heat to the weld area before interpass and during and following the welding process. This external application of heat treatment is referred to as interpass heating, preheating, and postweld heat treatment respectively. By minimizing the temperature differential during and after welding, the welded area will have lower residual stresses and is less susceptible to cracking and other metallurgical problems such as hydrogen damage.

The WELDHEAT system uses the same decision making procedure that an expert metallurgical engineer uses to establish weld heating schedules. However WELDHEAT provides a fast and efficient procedure to evaluate the heating requirements using several different methods. The system will interact with the user to choose the best analysis method (WRC recommendation, carbon equivalent, Pcm, or PHa) to use for preheating determination. Then the expert system will be called upon to assist with generation of a WPS and to verify that a standard or a developed WPS has the appropriate choice of heating schedules.

The database system incorporated with WELDHEAT contains "typical" composition values for over 500 ASTM classified steels. WELDHEAT will include in the decision process one or more of several important parameters, depending on the specific situation: cooling rate, potential hydrogen content, joint type, plate thickness, energy input, and electrode choice. The various methods can run in parallel. Based on the user's selection, one or more of the methods can be used to provide a "best estimate" of the preheat and interpass temperature, as well as the recommended postweld heat treatment. If the user or the blackboard does not have information about all of these parameters, WELDHEAT will use all of the available information to provide a "best estimate" of the preheat and interpass temperature as well as the recommended postweld heat treatment.

<u>Weld Procedure Specifier</u>

The objective of the Procedure Specifier Sub-blackboard (PSS) is to obtain a welding procedure which can be used to develop a weld schedule. That weld schedule is then passed to the welding job controller. There are three options which are available within PSS. First, if an applicable welding procedure specification (WPS) is available in the database which meets the requirements of the weld to be performed, then that WPS(s) is extracted. If there is more than one, the list is ranked in order of applicability and presented to the human user. Second, if no WPS is already available, then pre-existing welding procedure qualification records (PQRs) are extracted from the database which are applicable to the weld to be performed. Then a WPS is generated from these PQRs. Finally, if no applicable PQRs can be found, a PQR plan is developed which can be tested to produce PQRs in an actual weld testing operation.

<u>WELDPRO-*- Expert Systems</u>

This suite of expert systems is an important part of the WELDEXCELL WJP. Each expert system in the suite deals with weld procedure data. They work together to select an appropriate procedure (WPS) or to generate one, and then to develop a welding schedule based on the welding procedure data. The schedule is then used to direct the welding tasks to be performed.

The WELDPROSPEC expert system chooses a previously tested WPS from a database. If a WPS is not found which will meet the specific application needs, then WELDPROSPEC selects all of the PQRs from the database which are applicable to the weld to be performed. The PQRs which are selected are used to backup a WPS to be generated from the PQRs by WELDPROGEN. Additional rules are being added to the existing WELDPROSPEC system to include Mil. Spec. requirements; currently, it is based on the AWS D1.1 Code guidelines for WPS generation. PQRs are currently selected so that the WPS data specified falls within the allowed variance of the PQR essential variables as specified by AWS D1.1 code. The selected PQR data is passed to the WELDPROGEN expert system which is then called upon to generate a WPS. If appropriate PQRs cannot be located, the WELDPROPLAN expert system is called upon to develop a PQR test plan. WELDPROGEN will generate a WPS which conforms to the appropriate rules or guidelines from the set of PQRs which were selected by WELDPROSPEC. The expert system produces a WPS which contains all of the necessary data required to develop a weld schedule and which confirms to the applicable code or Mil. Spec. and which is adequately "backed up" by PQR's.

WELDPROSCHED develops a suitable welding schedule which is supplied to a manual welder, an automatic welder, or to the robotic welding system. Specific values and allowable ranges are supplied to define machine settings during the welding operation. The schedule considers position, thickness, and joint design changes and is able to adjust for multiple passes.

The Path Planner System (PPS) includes the basic design implementation of a welding path from a CAD-based design of the part to be produced. The three aspects of the system are (1) the CAD system which will be compatible with many commercial CAD systems including the Navy CAD systems; (2) the path planner which takes the CAD drawing and plans the welding path; and (3) a robot welding graphic simulator which includes collision avoidance assistance.

A computer-aided design (CAD) system is used that works in the specific hardware and software environment defined by this project. Currently the system uses AutoCAD. Drawings of the assembly pieces are created in three dimensions, and details of the joining of these pieces will be included. The CAD system must have the ability to allow the engineer to select the path to be followed by the welding torch/end effector of the welding robot. In addition, CAD code is being developed which allows the human user to identify objects in the CAD drawing. The three types of objects to be identified are (1) the weld line, (2) the objects to be joined, and (3) other objects which are potential collision objects or are important to the weld process (e.g. fixturing).

A separate system to be operated with the Welding Job Planner blackboard is available to do the path planning so that the path will be able to be associated with welding schedule data provided by the WELDPROSCHED system. Considerations are made for part accessibility, and/or possible obstructions on the part itself. It will also be necessary to simulate the robot movement relative to the parts to be welded. The simulation assists in path planning and collision avoidance.

A robot simulator has been developed for an articulated arm and for a gantry robot system. The simulator is capable of reproducing all of the robot motions including operating envelope limitations. The robot end effector world coordinates, and joint positions are displayed in real-time on the graphics screen. A specially designed collision avoidance system was developed by the Colorado School of Mines to operate in parallel with the real-time robot simulator.

WELDING JOB CONTROLLER

Functional Description

The Welding Job Controller (WJC) is responsible for all real time activities within the WELDEXCELL system. The WJC can accept a weld description from the Weld Job Planner (WJP) and accordingly control the welding hardware. The WJC will also collect data during the welding process for analysis on the WJP workstation. Figure 3 shows the top level organization of the WJC software components.

The WJC operator interface supports direct interaction between the end user and the welding system. Animated graphical control panels allow the operator to configure the hardware, adjust system parameters, load weld descriptions, and monitor the real time

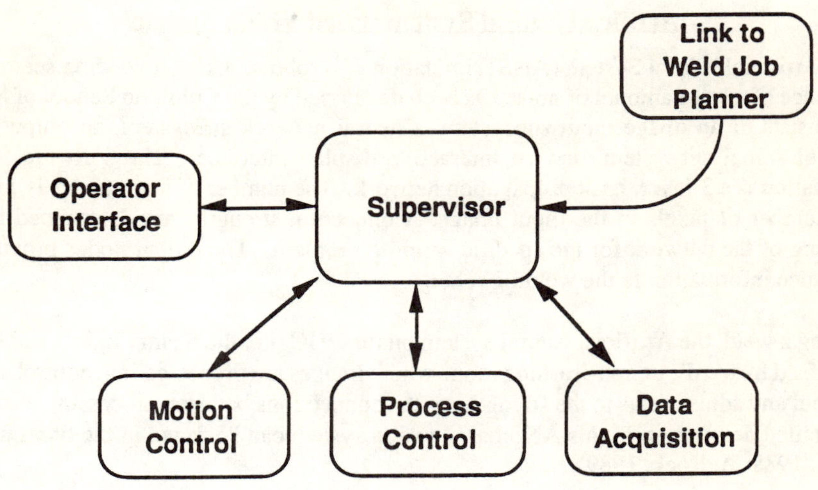

Figure 3. Overall Welding Job Controller Software Breakdown

welding process. The WJC system is currently under development by the WELDEXCELL team at the MTS Systems facility in Minneapolis.

COMPUTER SYSTEM

The WJP will be implemented on a Texas Instruments Explorer Artificial Intelligence Workstation, which is an ideal choice for fast execution of the various expert systems and databases which make up the WJP. An Ethernet link will be used for transferring files to and from the WJC. The WJC will be implemented on a VME-based Unix system consisting of several Motorola 680X0 microprocessors. This architecture is a proven platform for real-time welding workcell control and provides the openness which will be required to interface to different manipulators, welding equipment and process parameters.

<u>Welding Job Controller/Planner Interface</u>

The interface between the WJC and WJP has specific requirements that are primarily driven by the needs of the WJC. This interface is exclusively one of file transfer through a TCP/IP protocol LAN. The two primary activities of the interface are to pass weld schedule data from the WJP to the WJC and second to provide interpass weld history data to the WJP for interpass intelligent update of the process variables.

In order to provide maximum flexibility and modularity, the system was divided into two component subsystems at a point where minimum communication was necessary. This design provides for enhanced throughput and autonomy of operation. It also lends itself to a simple broad band LAN rather than to a bus structure. Finally, by subdividing the problem at this point, the engineering workstation can be remote to the actual workcell environment.

Artificial Neural System Based Vision System

An Artificial Neural System (ANS) simulation of a robot tracking a welding seam in the presence of a large amount of noise has been developed by the Colorado School of Mines. It consists of an image input subsystem, a neural network subsystem, an output robot control signal subsystem, and an interactive display interface. The software for the simulation is a 3-layer, back propagation network. The number of input nodes is equal to the number of pixels in the input image. Connection strengths are determined by the training of the network for the specific welding problem. The output nodes provide the guidance information to the welding robot.

During a weld, the Artificial Neural system on the WJC (i.e. the Seam Finder) will be run in a "feed forward", or non-learning mode where images are processed into control signals without any adjustments to the strengths of the connections between processing elements. A detailed description of this ANS based vision system can be found in the literature (A. Rock, 1988; A. Rock, 1989)

Real-Time System Implementation Strategy

The real-time processing components of the Weld Job Controller will be implemented using HOSE, a tool for programming industrial control systems. HOSE allows the designer/implementer to draw data flow diagrams that are automatically implemented on the real time hardware. The diagrams may be used for system design, simulation, implementation and diagnostics.

HOSE is used for both rapid prototyping and final implementation of industrial control systems at MTS. The graphical interactive nature of HOSE allows client feedback to be incorporated in the control system design process. Most of the diagrams used to illustrate the WJC real time systems in the previous sections are suitable for direct representation in HOSE. In many cases, the top level diagrams in a HOSE program *are* the design documentation.

Welding Control

The welding control system is designed to provide two capabilities. First, the system can set and hold a constant welding parameter schedule as specified by the WJP Subsystem. Second, the controller can ramp the welding parameters between physical set points in the welding path -- also as specified by the WJP. The welding control system provides the welding operator with the capability to view and, to a limited extent, to edit the weld schedule.

WJP User Interface Implementation Strategy

The interface will be developed through the use of the windowing facility in the Knowledge Engineering Environment (KEE) software and will interact with the X-windows software interface to provide easy accessibility to the user. If appropriate, the AllTalk language software, from MTS Systems, will be utilized. In addition, a significant effort is being made to unify the basic "feel" of the WJC and WJP user interfaces. This is not always possible, however, since the two interfaces require significantly different functions for different classes of users.

CONCLUSION

Based on the development of the prototype blackboard, expert systems, and databases for this project, it can be concluded that the ability to combine welding expert systems and databases is technologically very feasible. It has been estimated that the potential savings for the shipbuilding industry alone could be quite substantial if this technology is integrated into the welding activities of shipyards. Finally, it is the intent of the AWI/MTS/CSM team to complete the development of this system and to transfer this technology both to the ship building industry as well as other welding intensive industries in the United States.

ACKNOWLEDGEMENTS

The authors wish to express their appreciation to the Navy Manufacturing Technology Program and to Messrs. Steve Linder and Bob Achenbach for their financial support and encouragement during the first year of this program. In addition, the continuing support and interest of Mr. H. Brown Wright of the Tennessee Valley Authority is also acknowledged.

BIBLIOGRAPHY

1. S.G. Wax. et al., "The Intelligent Processing of Advanced Materials." presented at the 1986 TMS/AIME fall meeting, Orlando, Florida.

2. R. Mehrabian, "Process Control Schemes for the Advanced Processing of Materials," ibid.

3. B.G. Kushner and P. A. Parrish, "A Knowledge Acquisition Tool for the Intelligent Processing of Advanced Materials," presented at the 1986 TMS/AIME fall meeting in Orlando, Florida.

4. L.G. Tornatzky, et al., "Process of Technological Innovation: Reviewing the Literature," National Science Foundation, May 1983.

5. D.B. McCorkendal, "An Examination of Technology Transfer as a Tool for Management," Thesis, Naval Post Graduate School, March 1986.

6. R. Nigam and C.S.G. Lee, "A Multiprocessor-Based Controller for the Control of a Mechanical Manipulator," IEEE International Conference on Robotics and Automation, March 1985.

7. F. Ozguner and M.L. Kao, "A Reconfigurable Multiprocessor Architecture for Reliable Control of Robotic Systems," ibid.

8. M.D. Rychener, "High Level Tools for Engineering Design," preprints, OSU-DARPA Workshop on High Level Tools for Knowledge Based Systems," October 1986.

9. M.D. Rychener, "Expert Systems for Engineering Design," Expert Systems, 2.1., January 1985, pp. 30-44.

10. CTRL-C, *A Language for the Computer-Aided Design of Multi-variable Control Systems, User's Guide*, Systems Control Technology, Palo Alto, California.

11. *Hierarchical Object Structured Environment (HOSE)*, Reference Manual, MTS Systems, Minneapolis, Minnesota.

12. N.P. Nii, "Blackboard Systems: The Blackboard Model of Problems Solving and the Evolution of Blackboard Architectures," The AI Magazine, Summer 1986, pp. 38-53.

13. B. Hayes-Roth, "A Blackboard Architecture for Control," Artificial Intelligence, Vol. 26, 1985, pp. 251-321.

14. A Newell, "Some Problems of Basic Organization in Problem Solving Programs," Proc. Conference on Self Organizing Systems, ed. Yovits, Jacobi, and Goldstein, Spartan Books, 1962, pp. 393-423.

15. L.D. Erman, et al., "The Hearsay-II Speech Understanding System: Integrating Knowledge to Resolve Uncertainty," Computing Surveys, June 1980, pp. 213-253.

16. M. Rychener, et al., "A Rule-Based Blackboard Kernel System: Some Principles in Design," Engineering Design Research Center Report DRC-05-04-84. Carnegie-Mellon University, Pittsburgh, Pennsylvania, December 1984.

17. Technical Guide to the Welding Information Network, American Welding Institute, Knoxville, Tennessee, November 1986.

18. P. Oberly, et al., "An Artificial Intelligence Approach for Developing Welding Symbols," Presented AWS Annual Meeting, Chicago, Illinois, April 1987, to be published.

19. T.L. Funk, et al., "An Artificial Intelligence Approach for Selecting Welding Electrodes," Presented AWS Annual Meeting, Chicago, Illinois, April 1987, to be published.

20. "Technology Transfer Implementation Strategies for Intelligent Processing of Advanced Materials," International Conference on Manufacturing Science and Technology of the Future, '87, to be published.

21. R.W. Peterson, "Object-Oriented Data Base Design," AI Expert, Vol. 2, No. 3, pp. 26-31.

22. M. Athans, "A Tutorial on the LOG/.LTR Method," Proc. American Control Conference, Seattle, Washington, June 1986.

23. I.D. Craig, "The Ariadne-1 Blackboard System." The Computer Journal, Vol. 29,. No. 3, 1986, pp. 235-240.

24. A. Rock, et al., "Neural Network Applications in Automated Visual Weld Seam Tracking, Presented at the American welding Society National Convention, Washington, D.C., April 1989, to be published.

25. J. Jones, et al., "Development of an Offline Computerized Weld Planning System", Presented at the American welding Society National Convention, Washington, D.C., April 1989, to be published.

26. A. Rock, et al., "Investigation of an Artificial Neural System for a Computerized Welding Vision System", Proc. ASM Trends in Welding Research Conf., Gatlinburg, TN, May 1989.

27. A. Rock, et al., "Use of an Accelerated Learning Neural Network for Visual Weld Seam Tracking", Presented at International Neural Network Society Annual Meeting, Boston, September 1988.

28. H. Vanderveldt, et al., "Combining Welding Expert Systems with Welding Databases to Improve Shipbuilding Production", Proc. 1989 NSRP Ship Production Symposium, Arlington, VA, September 1989.

ALLOY MELTING EXPERT SYSTEM

Robert G. Trimberger and Barry R. Hathaway*
G.E. Corporate Research & Development
P.O. Box 8, K1-MB-139,Schenectady, New York 12301

*G.E. Consulting Services, Corporate Research & Development
P.O. Box 8, K1-4C9A, Schenectady, NY 12301

Abstract

General Electric's Corporate Research and Development Center melts over 1200 experimental alloys a year. The Alloy Melting Expert System (AMES) is a knowledge-based expert system that is used to select the furnace, crucible, mold(s), and raw materials and to generate specific melting instructions for each alloy. Mold, furnace, and crucible selection are each based on a number of factors such as reactivity, thermal characteristics, physical chemistry, and experience with similar type alloys. After evaluating all of the factors an ordered list is generated with the candidates molds, furnaces, and crucibles ranked according to suitability. Raw materials are selected primarily on the basis of purity, cost, volatility, and reactivity. Melting instructions are generated based on deoxidation requirements, alloy composition, type and form of the raw materials and vapor pressure of the raw materials. AMES consists of nine modules and five databases.

Expert System Applications in
Materials Processing and Manufacturing
Edited by M.Y. Demeri
The Minerals, Metals & Materials Society, 1989

Today's rapidly changing technologies require that new materials be created for use in new or existing products. In the field of metallurgy this means fabricating new alloys with properties different from alloys currently in existence. At General Electric's Corporate Research and Development (CRD) over 1200 experimental alloys are cast each year in quantities ranging from 25 grams to 250 pounds. These alloys range from materials with better corrosion-resistant properties to alloys intended to lengthen the life of aircraft engines.

Although on the surface the melting process appears to be a rather simple operation it actually involves a series of very complex reactions between the various raw materials that are to be melted. In addition to the interaction of raw materials there are also reactions between the crucible and the raw materials and between the atmosphere and the raw materials. The master melter knows what can be melted initially and what must be added later in the heat because it is volatile or reactive. He knows how to add very light or very heavy elements to the heat and insure that they will be dissolved properly; and he also knows how to handle the liquid metal to obtain the desired properties, taking into account such factors as whether the solidified material will be brittle and whether it will expand or contract on freezing.

There are a number of variables that must be considered in determining the proper procedures to use for each heat. These include:

• Mold material, size, and shape
• Furnace in which the melt will take place
• Crucible material and size
• Raw materials
• Temperature and pressure at various stages in the process
• Sequence of operations

These factors are determined both from scientific principles and from experience in melting similar alloys. This Expert System combines these scientific principles and the experience of several experts.

System Concept

The concept behind the Alloy Melting Expert System (AMES) was to develop a system that given a chemical composition, it would be able to generate a set of detailed melting instructions, identifying the proper furnace and crucible to use, as well as selecting the raw materials. Additionally it was desired that the AMES system also provide:

• The ability for an operator to override any decision that the system may make, if necessary
• Explanations of the reasoning used in making decisions
• An easy-to-use interface so that people with little or no computer experience could make effective use of the system
• The ability to add, delete, or modify rules as experience is gained in melting new alloys.
• Easy maintenance of the system databases

• On-line documentation and a HELP facility

The resulting system is composed of nine separate modules: customer data input, composition input, mold selection, furnace selection, crucible selection, raw materials selection, loss calculation, melting instructions generation, and output (Figure 1). The mold selection, furnace selection, crucible selection, raw materials selection, and the melting instructions generation modules are rule-based expert systems. Communication with one another is via a blackboard approach. The system also makes use of a number of databases containing information on materials and equipment on hand. These include information on chemical elements, molds, furnaces, crucibles, raw materials, and losses incurred in the melting operation.

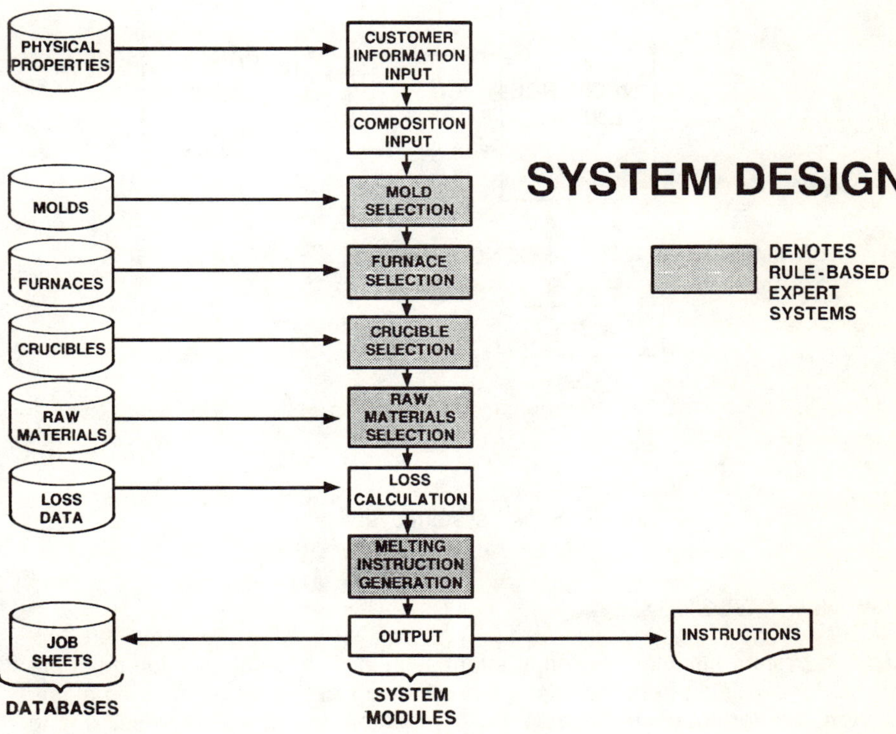

Figure 1. System design

The various selection and generation procedures required the capturing of various types of knowledge. The knowledge ranged from information on physical chemistry to experience gained only through making similar alloys in the past. To capture all this information, standard rule-based expert systems were constructed consisting of a rule base, a fact base, and an inference engine (Figure 2). Because of the nature of the rules and the selection processes a forward-chaining approach was chosen. Here the inference engine tries all available rules over and over, adding new facts to the fact base, until no rule applies.

EXPERT SYSTEM OVERVIEW

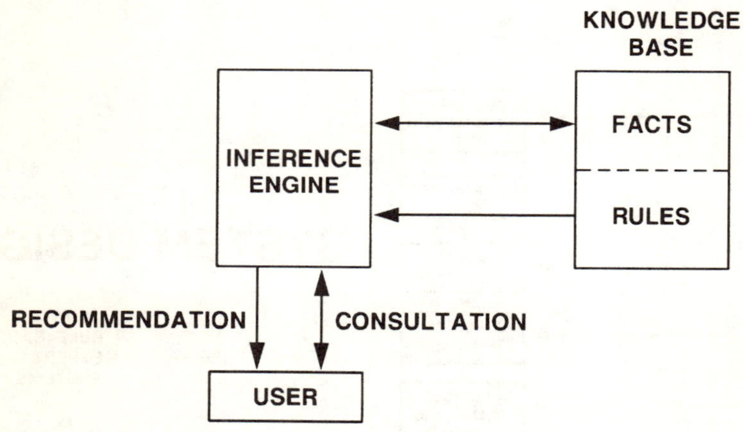

Figure 2. Expert system overview

<u>Customer and Composition Input</u>

The first two system modules, customer information input and composition input are responsible for obtaining from the user information concerning the alloy to be made. The customer information input module, in addition to name, address, phone number, and shop order number, allows the selection of a cooling curve and/or a low sulfur heat. If a cooling curve is desired a change in the melting instructions is triggered. A request for a low sulfur heat results in changes in material selection and melting instructions. The composition of the alloy can be given in either weigh or atomic percent. The system automatically calculates the other. Alloy composition can also include compounds. Figures 3 and 4 show the customer input and composition input screens.

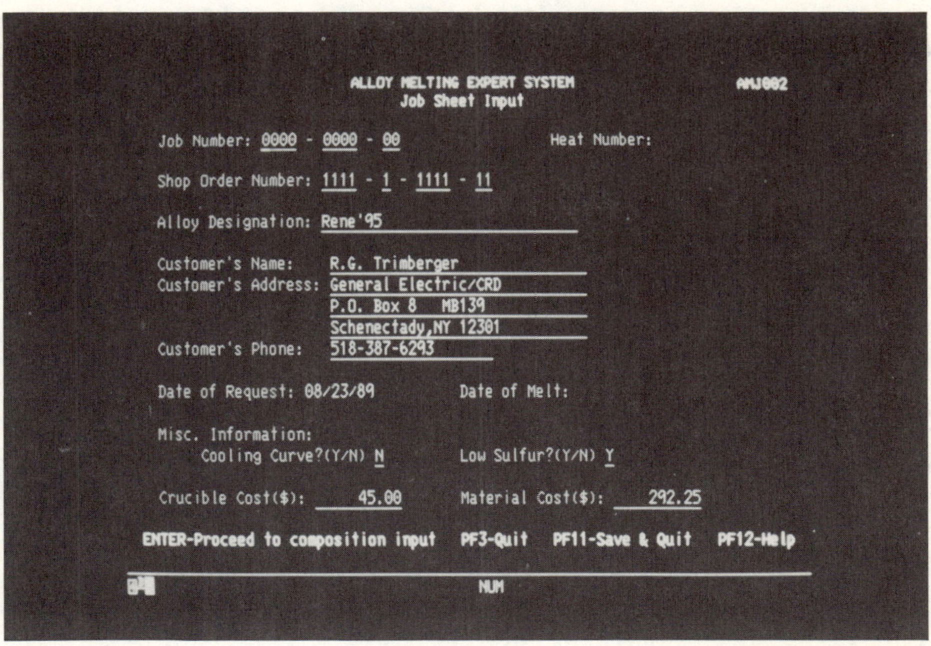

Figure 3. Customer information input

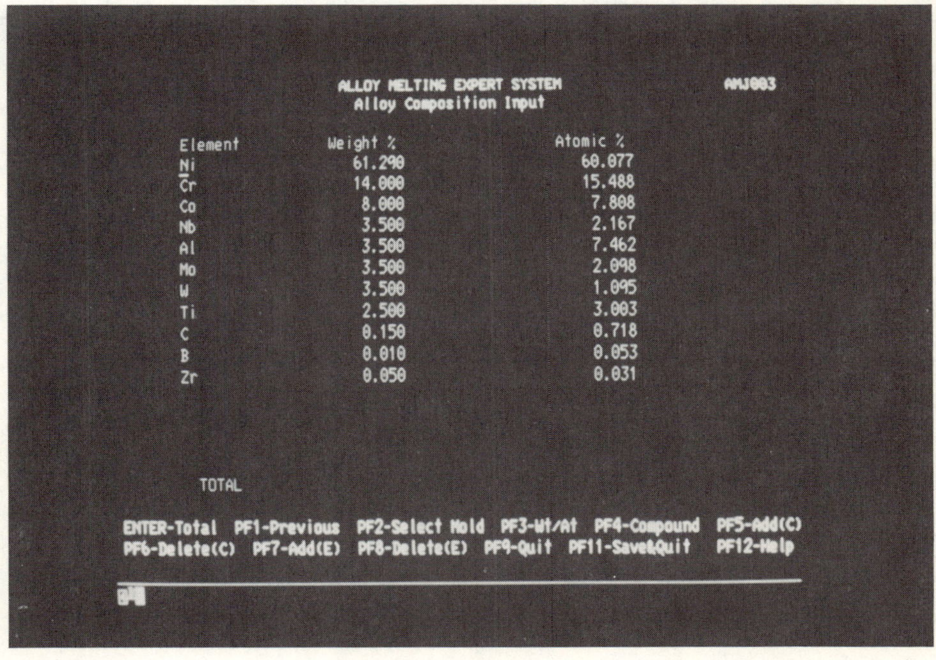

Figure 4. Composition input

Mold Selection

The mold selection module is used to pick the proper mold in which to pour the molten metal. The mold can be selected from seven different types, six different materials, and an almost infinite variety of sizes. Although it can operate in the automatic selection mode normal practice is to select a mold manually.

In order to ascertain the proper mold it is necessary to sit down with the customer and determine what his end requirements are for the ingot, and how much material is required. The soundness of the ingot must also be considered. By the time the discussion with the customer is finished the mold characteristics have been well defined and the proper mold has been selected. In the automatic mode these same factors are considered in a rule-based expert system. Using the rules and the database of available molds, the system comes up with a set of acceptable molds ranked according to suitability on a scale from0 to 1. The user can then request an explanation of any mold ranking and select any mold on the list.

Furnace Selection

Some of the important factors that must be considered in selecting a furnace include:

- Alloy composition
- Pressure requirements
- Heat size
- Deoxidation requirements
- Availability of furnace and operator

Heats that contain large quantities of reactive elements such as Titanium should not be run in furnaces that require the use of oxide crucibles because the reactive element will reduce the crucible material leading at best to high oxygen contents and at worst to a total failure of the crucible resulting in extensive damage to the furnace and possibly injury to the operator. This limits the melting technique that can be used to some type of a cold crucible system. High vapor pressure materials, on the other hand, require a furnace that can be pressurized to restrict the losses of the volatile material.

To accomplish the furnace selection, the system employs a rule-based approach, forward chaining through a set of generic rules and a set of site-specific rules. The generic rules are used to indicate the melting technique to be used, pressure requirements, and other factors, thus narrowing down the list of possible candidates. The site-specific rules finalize the decision based upon experience gained in using the furnaces. The final result is an ordered list of candidates ranked according to suitability with each furnace receiving a score in the range of 0 to 1 (Figure 5 and 6).

```
                    ALLOY MELTING EXPERT SYSTEM              AMJ030
                       Furnace Selection

Furnace Name: BPV                        . Heat size (pounds):    29.706


                    List of Installed Furnaces

                 Capacity Range(lbs)   Pressure(psi)  Deoxidation
Furnace Name       Low      High  .    Low   High    Capabilities    Score
BPV              10.000    55.000     0.0  164.7   H2    C-H2         1.00
CVC              25.000   300.000     0.0   16.7   H2    C-H2         0.90
ARC               2.000    30.000     0.0   15.0                     0.25
1 lb              0.000     4.000     0.0   14.7                     0.00
Retech            0.100     5.000     0.0   14.7                     0.00
PV                0.000     5.000     0.0  164.7                     0.00
Non-Consumable    0.500     2.000     0.0   14.7                     0.00
Multi_Hearth      0.011     0.442    14.7   14.7                     0.00
LPV               5.000    12.000     0.0   74.7                     0.00
Consumable Arc    5.000    20.000     0.0    0.0                     0.00


    PF1-Prev. Screen  PF2-Analyze  PF3-Explain Score  PF7-Backward Furnaces
    PF8-Forward Furnaces  PF11-Save & Quit  PF12-Help  ENTER-Select Furnace
```

Figure 5. Furnace selection

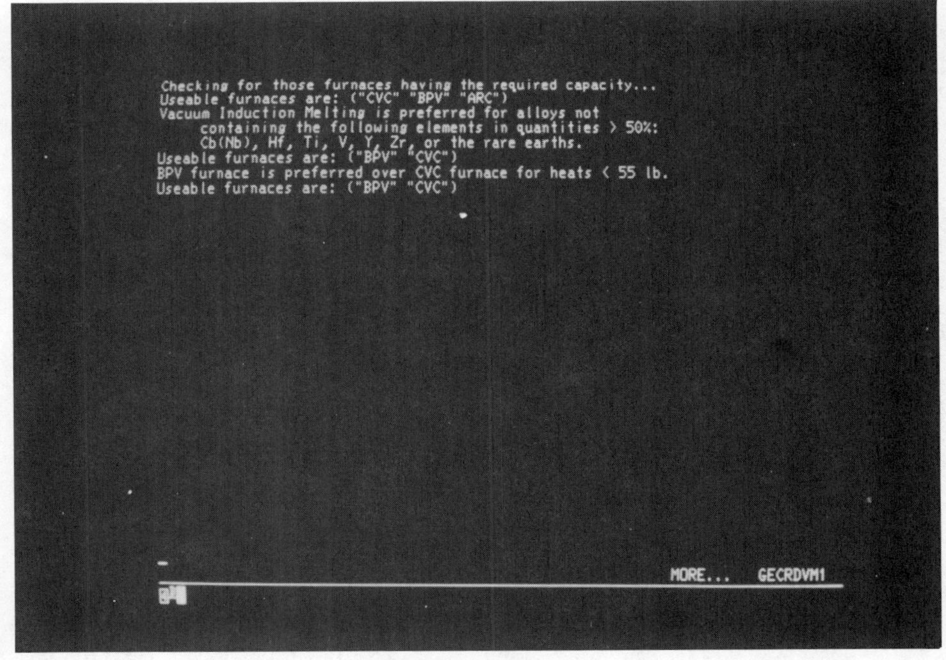

Figure 6. Explanation of furnace selection

Crucible Selection

Furnace selection is followed by choosing a crucible in which to melt the raw materials. Crucible selection is one of the most critical aspects of the entire melting operation. As mentioned in the furnace selection section the choice of the wrong crucible material can lead to a total failure of the system, possibly resulting in a catastrophe. Determining the proper crucible involves choosing both the right material (water-cooled copper, alumina, magnesia, zirconia, graphite, quartz, or other specialized materials) and the right size. Factors involved in crucible selection include:

- Alloy composition
- Melting points of individual raw materials and the alloy
- Reactivity of materials with crucible
- Volume
- Dimensions
- Furnace
- Experience in melting similar alloys

Water-cooled copper is the best choice for melting highly reactive materials and very-high-purity materials, but it has the disadvantage of requiring high power inputs to overcome the chilling tendency of the water-cooling. It is very difficult to obtain a homogeneous alloy in this type of crucible because the entire heat cannot be kept molten at the same time, so each heat must be remelted a number of times. The oxide crucibles can maintain the entire heat molten at one time, so they will normally make a more homogenous alloy in the liquid state (the solidification parameters in the mold determine the final homogeneity of the solid), but they have the disadvantage of reacting with a number of materials. One measure of the tendency of the oxides to react is their free energy of formation. The table below ranks some of the oxide crucibles in order of increasing stability (based on free energy of formation):

Table I - Oxide Crucibles Ranked in Increasing Stability

SiO_2
MgO
Al_2O_3
ZrO_2
BeO

Although beryllia and thoria are the most stable of the oxide crucibles, they are very seldom used because beryllia is toxic and thoria is slightly radioactive. As a general rule SiO_2 crucibles are not used except for melting silicon-based heats. The choice of the proper crucible material is based primarily on the alloy composition, paying particular attention to the reactivity of the various elements with the crucible material and also the maximum temperature that the molten metal is expected to reach.

There are also a number of situations where the normal rules dictate the use of a particular type of crucible but experience has shown that another type will perform better, so there are some exceptions to the rules that are considered when selecting the proper crucible.

A rule-based expert system is the heart of the crucible selection procedure. Two distinct rule sets are used. The first set is used to rank available crucible materials according to suitability. This is followed by the use of the second set of rules to determine the size of the crucible required. The result of the selection process is an ordering, based upon a score from 0 to 1, of all crucibles that will physically fit in the selected furnace according to their perceived suitability . At that point the user can either choose the crucible with the highest score or override the systems choice and pick any crucible. Explanations of any crucible's ranking can also be obtained.

In addition to picking the crucible, this module also determines whether or not a susceptor is required to assist in the melting operation. Susceptors are required when the charge is non-conductive or when the volume of the heat is too small to allow the field to be picked up by the material in the crucible. If a susceptor is required, the melting instructions will so indicate.

<u>Raw Material Selection</u>

Raw material selection follows crucible selection. This involves choosing the individual metallic elements that will make up the alloy and calculating their proper weights. Consideration must be given to whether the material will fit in the crucible, how to add the material, how to weigh the material accurately, how it will react with the other elements, and many other factors, In many cases master alloys must be used to overcome the disadvantages that some materials pose. Some of the advantages that can be gained by the use of master alloys are listed below.

• Increased ease and accuracy of weighing very small quantities.
• Decreased possibility of losing very small addition samples while adding
  and during the melting process.
• Lower vapor pressure than the pure element.
• Lower or higher melting point than the pure element (as desired).
• Lower or higher density than the pure element (as desired).
• Lower reactivity than the pure element.
• Custom tailoring to fit a particular problem.

To accomplish the task of raw material selection, the Raw Materials Selection module uses a rule-based expert system that considers such factors as:

• Alloy composition
• Reactivity of materials with other materials
• Deoxidation requirements
• Purity
• Stability
• Furnace

- Vapor pressure of the various materials
- Gas content
- Form and shape
- Whether the material is a single element or a master alloy
- Availability

Using the alloy composition and the database of raw materials, the expert component can select some of the materials necessary to fabricate the alloy. In general, after the rules have been exhausted, all materials necessary still have not been selected. This is because the rules in the rule base do not cover all the possible raw materials. At that point any material containing the remaining individual elements are selected, considering only purity requirements. Throughout the selection procedure if a material meeting the purity requirements of the alloy is not found than the system will attempt to substitute a material of a different purity grade. The same is true if the supply of a material has been exhausted, In all cases the user is informed of any substitutions (Figures 7 and 8).

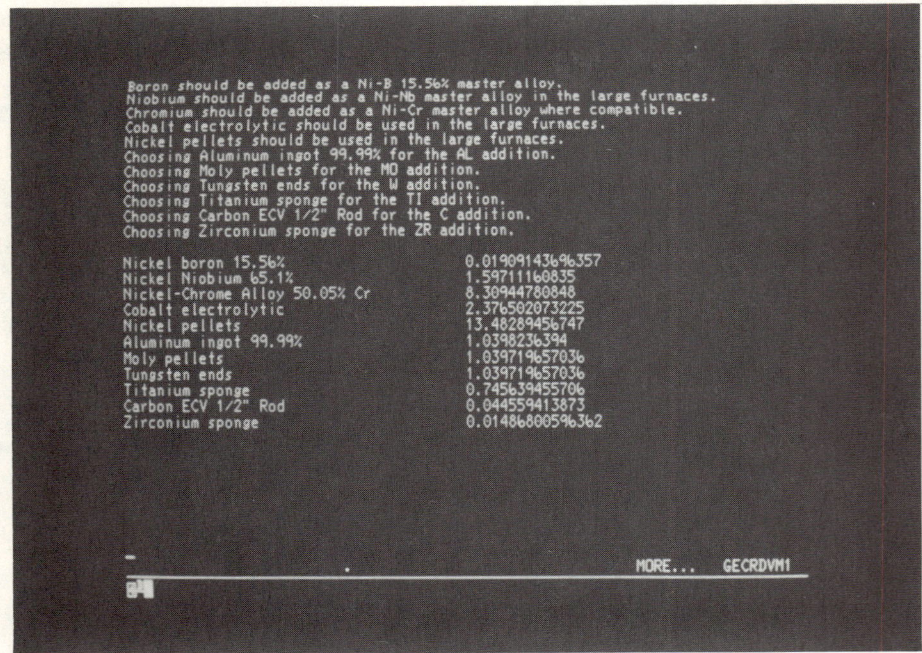

Figure 7. Raw materials being selected

Loss Calculation

The raw material quantities determined above must then be adjusted by the Loss Calculation Module to compensate for losses that will occur in the melting process. This can be compensated for by adding more of the raw material   The system maintains a file of those materials that will incur losses along with typical loss percentages. Using the file, the Loss Calculation Module makes the proper adjustments. The user has the option of overriding any adjustments and can also vary the recovery rates used in the calculations.

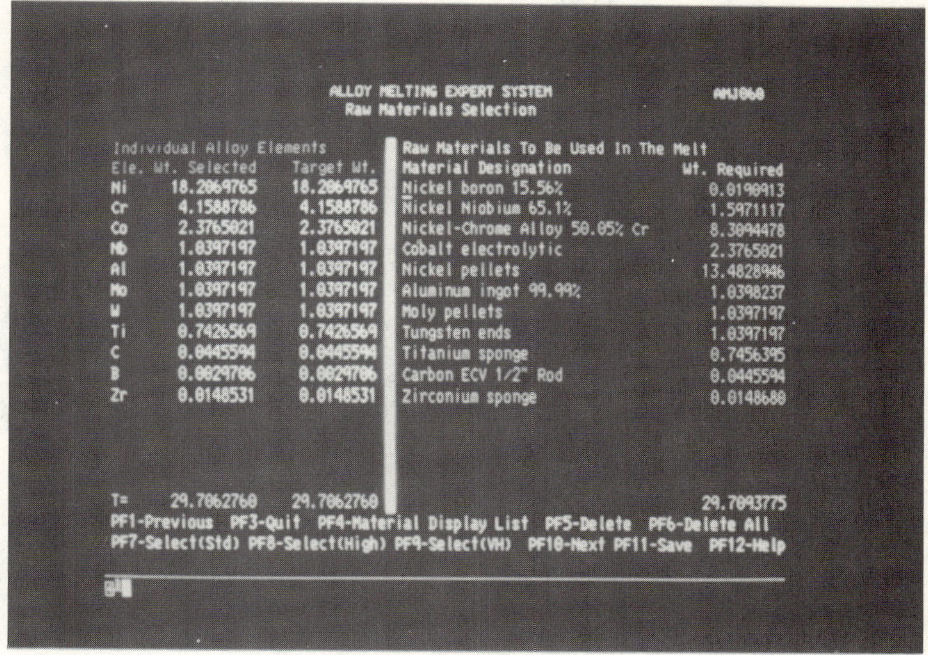

Figure 8. After raw materials selection

Melt Instruction Generation

The generation of melting instructions follows the selection of equipment and materials. This task involves specifying how the equipment is to be used and the order of adding the raw materials. The melting instructions vary greatly between different alloys calling for consideration of such items as:

• Initial charge composition
• Type of atmosphere (vacuum, argon, hydrogen, etc.)
• Pressure
• Gas flow rates
• Melt and superheat temperature
• Time
• Melt-solidify sequences
• Order of additions
• Pour temperature and pressure

Such a large number of variations in the melting instructions requires the Melting Instruction generation Module to consider an equally large number of factors. These include:

- Furnace
- Heat size
- Alloy composition
- Type and form of raw materials
- Purity requirements
- Gas content
- Vapor pressure of materials
- Reactivity of materials with each other and the crucible
- Presence or absence of nitrogen in the alloy.

Of critical importance are the specification of the initial charge (i.e., the materials that are melted together first) and the order of adding the remaining materials. If this step is not carried out correctly, materials may react with each other or with the crucible, certain elements may incur high losses because of vaporization or oxidation, or elements may not dissolve completely in the molten charge. Rules for determining the above turn out to be fairly complex and vary greatly depending on the alloy composition. For example: copper can be part of the initial charge for aluminum-based alloys, but not for nickel-based alloys.

To take into account rules of the form mentioned above, the Melting Instruction Generation Module had to employ a rule-based expert subsystem. By forward chaining through a single set of rules, the initial charge makeup can be easily determined. Subsequently the order of any other late addition can be fixed.

Determining some of the other variables, such as atmosphere, gas flow rate, and sequences of operations are pretty much dependent on what furnace is used in the melt. Here the furnaces can be grouped into sets, and a traditional programming approach (in LISP) can be taken to generating the remaining melting instructions for each set (Figure 9). The pressure requirements are primarily determined by alloy composition and the maximum temperature of the melt.

A notable exception is the case of creating nitrogen-bearing stainless steel alloys. These alloys contain quantities of nitrogen, usually less that 0.3% by weight, dissolved in the metal. The only acceptable way of accomplishing this is to add the required amount of nitrogen as a component of a master alloy and pressurize the furnace using a nitrogen-argon atmosphere. By choosing proper partial pressure of nitrogen in the atmosphere, the nitrogen dissolved in the molten liquid can be maintained at the proper level. The partial pressure is dependent upon the amount of nitrogen to be dissolved and the other elements in the heat and must be calculated with great accuracy. This accuracy was achieved by running a series of test heats, adding a known quantity of nitrogen and varying the partial pressure for each heat. The resulting ingot was then analyzed for nitrogen content. The procedure was repeated for a number of "typical " alloys. The data was then plotted to yield a correspondence between the weight percent of nitrogen in the alloy and the square root of the partial pressure of nitrogen that must be applied

The Melting Instruction Generation Module is responsible for determining the correct partial pressure to use for a given Nitrogen-bearing steel. This capability was achieved by fitting second-order and third-order polynomials to the data points for each "typical" alloy. Based upon the composition of the alloy to be created the system attempts to find which one of the "typical" alloys it most closely resembles. This requires the use of a special set of heuristics to carry out the matching process. Once the "typical" alloy is found, its polynomial is evaluated using the required nitrogen percentage to obtain the partial pressure specification.

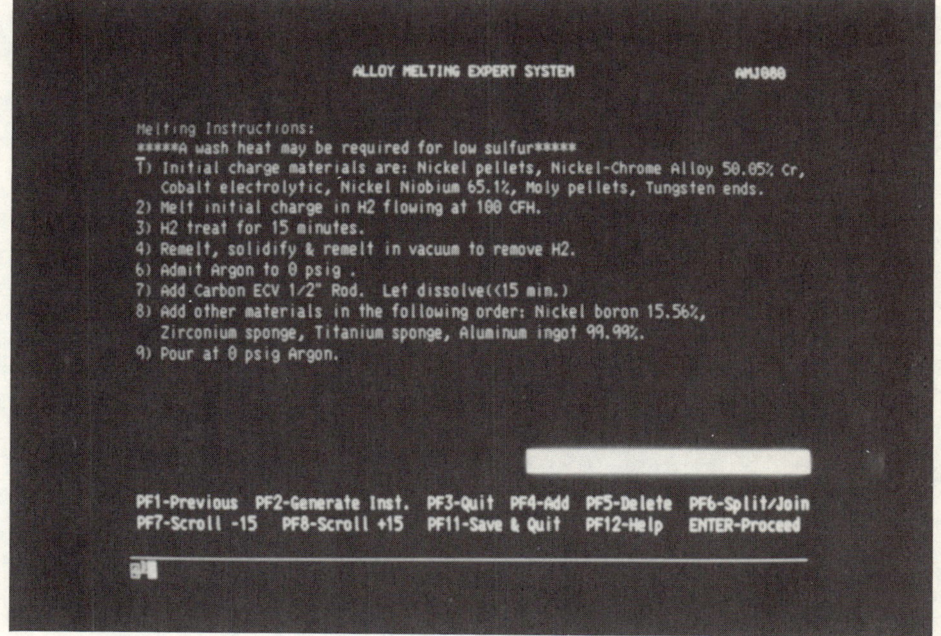

Figure 9. Melting instructions generation

Summary

The AMES system has been revised three times so far. The original version had a success rate of about 75%; ie, the materials and procedures selected by the system did not have to be manually changed in three out of four cases. Most of the changes that did have to be made were relatively minor. Since the melting process is a dynamic operation and since we are constantly venturing into uncharted areas in the search for new and better alloys there will probably never be a 100% success rate. New rules will have to be formulated to cover the new situations as they arise, but it is anticipated that the success rate will exceed 95%.

Developing an Expert System

for the Heat Treating Industry

Edward D. Jamieson and Terrence D. Brown

Lindberg Heat Treating Company
3601 West Algonquin Road
Rolling Meadows, Illinois 60008

Abstract

As computer based control and management systems become more integrated into heat treating operations, the development of expert systems becomes more feasible. While expert systems reside in all manufacturing operations with experienced personnel, the ability to access that information remains limited. Lindberg Heat Treating has begun to develop an expert system based on experienced personnel and process control signals. The system will be designed to be interactive with shop personnel, engineers and managers. The program will address the development of expert systems for real time process control based on computer modeling and process signals. Historical process parameter data will be inputted to the decision-tree from the computer based control unit to aid in decision making. Equipment maintenance and troubleshooting are also part of the program. In addition, a training program using tutorial techniques will be developed to aid shop personnel in utilizing the expert system.

Expert System Applications in
Materials Processing and Manufacturing
Edited by M.Y. Demeri
The Minerals, Metals & Materials Society, 1989

Introduction

When the heat treating of metals entered the twentieth century, it relied on the experience of skilled craftsmen. Their experience in developing heat treat processes was refined over the centuries by applying trial and error methodology. The period was characterized by a limited number of materials, wide tolerances and simple geometries. Simple "rules of thumb" were developed to address the limited number of process variables encountered.

The early part of this century saw a rapid growth in the metalworking industry spurred by the requirements of mass production and developments in the science of metallurgy. The number and types of metals and their possible application grew at a rate never encountered before. The heat treating industry also grew to meet the requirements of this changing market place. One processing feature that did not change in the heat treating industry was its reliance on experienced people (or experts). These individuals and their apprentices would continue to play the pivotal role for decades to come.

Background

Computers and microelectronics are probably the most important industrial developments of this century. However, the heat treating industry has lagged far behind other metalworking disciplines in their application. This slow development may be traced to the high reliance on experts and the multitude of variables involved in the process. As the number of experts declined and the complexity of heat treating increased, it became apparent that a computer aided system must be developed to capture the experience of those few remaining individuals.

Lindberg Heat Treating recognized this opportunity in their operations and began a program to address it in the early 1980's. The program was organized into three phases. The first phase addressed the business side of the operation. This included such programs as order writing, order tracking, payroll, accounting, etc. The next phase targeted the heat treat process, controlling the operation of all equipment via a computer to control and/or monitor all process variables. The final phase combined the two systems into an integrated plant management system. It is in the final phase that will create the foundation of an expert system.

While the business side of the computer system was fairly straight-forward, the second phase proved to be much more challenging. In order to understand the direction needed for developing a computer system for the heat treat process, one must examine the four major factors that affect any heat treating process. They are:

> 1) Material
> 2) People
> 3) Equipment
> 4) Methods

Material

The material provides the direction the heat treat process will take. It will define the limitations that impact the equipment selection and heat treat methods. Although, the heat treater has little control over the material. But any information or data about the material is useful in predicting results. Therefore, a provision had to be made to input and manipulate this information. The purpose of this was to develop a knowledge base for heat treating control parameter recipes.

People

The people provide the intangible element to the process. While people have been needed to insure that all process parameters are set properly, they are also required to adjust the process according to loading patterns, geometries, etc. The furnace control system was targeted to eliminate the demand on the operators to set and adjust all process variables within the correct tolerances throughout the process.

136

<u>Equipment</u>

No two furnaces are identical in the way they operate. Provisions within the system were designed to adequately monitor the equipment in order to know when it is operating out of tolerance. Out of tolerance conditions must be recognized before an out-of-spec product is produced.

<u>Methods</u>

The method of developing heat treating recipes has not changed significantly from the historical trial and error method. A heat treat recipe for a particular component evolves from this method. Repeatability of the given recipe insures the predictability of the end results. Statistical analysis of the product characteristics and the process variables can be used to optimize the recipe and the control of the process.

It is the interaction of these four (4) factors that defines the heat treating process. While the control and/or monitoring of the major process variables within each factor was relatively simple, the sensing of secondary variables proved very costly and difficult.

These difficulties were eventually overcome and the initial system of Lindberg's Computer Aided Heat Treating System (CAHTS) was completed in late 1988. The system design has a three level architecture: level one consists of local processing units (LPU's) which are shop hardened devices that control the furnace; level two is a management station which can be linked to 16 LPU's and is used to configure the control loops, process alarms, reporting, real time graphics and logic functions; and, level three, a host plant computer that provides recipes, data archiving, operator interface, bar coding, material tracking and networking capabilities. The levels one and two devices are manufactured by Leeds and Northrup and level three is a DEC Microvax computer. It is this system that will provide the foundation of the expert system (see Figure 1).

This paper depicts the strategies and approaches being utilized to enhance the CAHTS system with a knowledge base. The final objective will be the evolution of an expert system for heat treating.

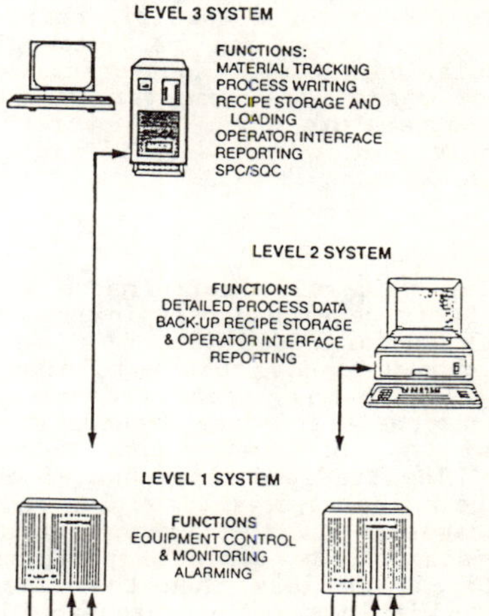

Figure 1 – Three level architecture which provides
the foundation of the knowledge base system.

Applications

The development and application of expert systems in the heat treating process have been organized into four areas:

1) Recipe Selection
2) Scheduling
3) Real Time Process Control
4) Equipment Troubleshooting and Maintenance

While these are not all the possible applications, they do represent the highest payback for the development time and cost involved.

Recipe Selection

The recipe selection process involves the translation of design engineering specifications to measurable product characteristics. The target product characteristics determine the treating process control parameters. An example of this would be the converting of a tensile strength range to a specific hardness range. It is the defining of these control parameters that offers the initial application of an expert system. Typical control parameters that would be defined are: austenitizing temperature, time at austenitizing temperature, quenching medium and tempering temperature and time.

The process and material information required to make these control parameter decisions comes from two sources; published metallurgical data and "rules of thumb". The "rules of thumb" come from the practical experience of experts. Some "rules of thumb" are: never water quench alloy steel parts that have cross section areas of less than one inch, always suspend parts vertically if the length/diameter ratio exceeds ten, if the application of the part required high impact and toughness use the lower end of the austenitizing range. As more of these "rules of thumb" are automated into the recipe selection process, the more "expert" the system becomes.

To complete the development of the expert recipe selection system, a feedback loop will be designed. This will provide a link between actual resultant product and the actual process control variables. By statically analyzing the historical data for a specific combination of part geometry, material and furnace, a recipe can be automatically fine-tuned. For instance, if a particular part number consistently tests at the high end of the specified hardness range, the computer will automatically raise the recipe's tempering temperature by $25^{\circ}F$, or if the case depth is consistently on the low side of the specification, the carburizing time will be increased by 30 minutes.

Scheduling

In most scheduling systems, a performance standard is defined and a method of monitoring it is developed. Typically, the standard has a 15 to 30 percent contingency factor added into it. This contingency factor is added and varies by furnace and process to insure the product quality. While the contingency factor may appear excessive, it is not because the number of variables that can affect the process are taken into account.

The minimizing of the contingency factor is the goal of the expert system in scheduling. The traditional method of obtaining feedback on the schedule is to have the operator fill out a log on the furnace performance. The log is used to categorize all the furnace hours. The log would then be compared against the schedule to evaluate performance. If, and only if, the actual time required was less than the time allotted, and all the furnace parameters were within a given range, would the standard change.

The data collection portion of the scheduling system is now being performed by the computer without the interaction of the operator. The standards which were defined by pounds per man-hour or pounds per furnace-hour tended to be generic and did not address the actual production time of each part or process. These generic standards have been replaced by standards based on the recipe key which has immediate access to the previous process history. The specific recipe data is being archived to allow for refinement of standards. This is currently a manual operation, but will soon be automatic.

138

The contingency factor is being reduced by two previously mentioned items. The two items are the development of specific schedules based on previous part history and the refinement of previously used scheduling standards. It is also being reduced by the ability to more precisely control and monitor the heat treating process itself. With each successful run, the opportunity presents itself to adjust the process parameters somewhat to increase production without affecting quality. This feature automatically comes up on specific process variables and prompts the order writer to seek approval to change the process parameters or leave them the same.

Modeling has been utilized to predict times of cycle segments that are not controlled. For instance, the time for a load of parts to reach austenitizing temperature can be estimated by a model that factors in gross weight, part geometry, load distribution (fixturing), process temperature and preheat temperature. The model can be refined via historical data. By accurately estimating the heat-up time for a load, the scheduling process becomes more precise.

Modeling can be used as a scheduling tool to optimize throughput of continuous furnaces. A carburizing model is used to test various combinations of times, temperature and atmosphere that will produce a specified carburizing case. This can be used to minimize gap times between change overs. As an example, the affect on case depth and carbon profile can be predicated based on several combinations of zone temperature push time and carbon potential in a continuous pusher furnace. By reviewing the available combinations, the optimum recipe can be selected that will optimize productivity. For instance, it may be better to lower the carburizing temperature rather than speed up the push time. To minimize the number of empty pushes between lots.

Real Time Process Control

Three different methods are being developed to provide real time process control. Firstly: utilizing a computer model of the process to assures it is proceeding under a specific set of conditions. Secondly, the heat treating process parameters are being monitored on a continuous basis to insure they do not deviate from a specified set of ranges. Lastly, the process parameters are being set by the computer and not the operator. The main objective is to provide information quickly so that the operator can respond to a process that is approaching an out-of-control situation, before out-of-spec parts are produced. The carburizing process has been the focus of several different programs to develop a computer based model of the process. The current models take various inputs such as material, temperature, carbon potential and case depth and predict a carbon gradient along with an expected time cycle. It is this type of model that has been modified to take actual process values of temperature and carbon potential, that will be utilized. The computer model will be used to compare the current process performance based on actual furnace conditions with its theoretical performance based on calculated or experienced driven values inputted into the model. It will then take actual furnace conditions into account and modify the ongoing cycle to guarantee the components will meet the required specifications at the completion of the cycle. The modification of the cycle can be programed to happen automatically or require operator approval. It will also be programed to alert the operator when the executing process deviates from the model. Following the completion of each cycle, the model derived from process results will be refined and fine-tuned.

The ability to monitor every variable on a continuous basis has allowed the control of the process to reach new heights. Each process variable (ie., furnace temperature, carbon potential, atmosphere, flow rates) is now given a tolerance range to operate within. When a variable deviates from the given range an alarm is triggered automatically. These alarms are documented along with the operator acknowledgement and corrective action.

The final method for real time control relates to the setting of actual furnace variables. While this is the least complex task of the control system it does impact most significantly on the process. The automatic changing as a function of cycle time assures the repeatability of the cycle. This function also allows the monitoring of the process control variables to be continuous throughout the process.

<u>Equipment Maintenance</u>

One of the first areas an expert system was utilized was on equipment breakdowns. Previously, a written instruction manual was developed on each unique piece of equipment to instruct maintenance personnel on how to repair what was defective. What the manual did not address in detail however, was how to determine what was defective. That identification has been addressed through the use of the CAHTS system. Because of the large number of process variables and furnace operating functions being monitored and alarmed, the system now has the ability to detect if something malfunctioned on the furnace. Once a malfunction occurs, a process variable or furnace operating function will go into alarm. When the operator or maintenance individual answers the alarm, a set of remedial actions are provided. The actions are also ranked in order of probable corrective action. Since a malfunction can manifest itself in a variety of different sensors, the ranking is calculated by assigning a value to each process variable sensor that is monitored and might be influenced by a malfunction. The values take into account whether the sensor is in an active alarm condition or within the pre-defined normal range. An example is that a defective radiant tube can be detected by the combination of; higher than normal furnace pressure and higher oxygen content of the furnace atmosphere or greater flows of methane additions required to maintain carbon potential.

Preventive maintenance steps are triggered via the computer. Lubrication schedules, sensor calibrations and replacements of high alloy heat resistant components are initiated by computer prompts. The actual maintenance records are also recorded in the CAHTS system.

<u>Future Considerations</u>

While the development of the aforementioned expert system applications covers a wide variety of situations, they do not address how to expose large number of individuals to the experience and intuition of the experts. To accomplish this we are developing computer-based tutorials to teach the problem solving strategies and principles inherent in the expert's knowledge.

The first tutorial lessons being developed are on carburizing. The basic principles of the process will first be presented. The next level of the carburizing tutorial will address the interaction of the process variables. The third level will consider the various industrial applications which utilize the carburizing process. The final section will review how improper carburizing will manifest itself in the various applications where it is utilized.

The applications presented in this paper represent a knowledge base system more than an expert system. The knowledge bases developed to date have been the practical experience of heat treating experts who imparted their knowledge through question and answer sessions. While the experience of these individuals has formed the foundation, they cannot be expected to build the rest of what is needed. It will only be through the refinement of this foundation, through the use of a continually updated data base, that heat treating will be able to have a truly expert system.

References

1.   Sholom M. Weiss, and Casimir A. Kulikowski, _A Practical Guide to Designing Expert Systems_ (New York, NY:Rowman & Allanheld Publishers, 1982), 1-16.

2.   Stephen M. Alessi, and Stanley R. Trollip, _Computer-Based Instruction Methods and Development_ (Englewood Cliffs, NJ:Prentice-Hall, Inc., 1985), 65-131.

3.   Wendy B. Rauch-Hindin, _Artificial Intelligence in Business, Science and Industry_ (Englewood Cliffs, NJ:Prentice-Hall, Inc.), 63-87.

4.   T.D. Brown, "Strategies for a Computer Aided Process System" (paper presented at the Gas Research Institute Workshop on Advanced Combustion and Process Control, Chicago, IL, 23 August 1989), 5.

5.   T.D. Brown, and E.D. Jamieson, "Computer Aided Heat Treating Makes Treatments Compatible with Production Environments," _Industrial Heating_, 5 (1989).

AN INTEGRATED ENVIRONMENT FOR INTELLIGENT DESIGN OF

CASTINGS

R. Natarajan, C. N. Chu,* and R. L. Kashyap,***

*School of Industrial Engineering
Purdue University
W. Lafayette, IN 47907

**Knowledge Based Systems Lab.
School of Electrical Engineering
Purdue University
W. Lafayette, IN 47907

Abstract

Casting process is heavily experience-oriented and casting design is an iterative task between casting designers and foundry experts. A knowledge-based expert system called EXCAST was developed in order to facilitate the design and the manufacturing of casting components. The EXCAST system uses a fixed-feature design approach, where a generic group of features is used to facilitate the design process. Tolerancing, rounding, shrinkage, simple junction, and wall thickness are incorporated in the rule base. These rules are suitable for evaluation of local casting features. However, casting components generally have complicated geometric shapes, and design rules cannot always address the solidification behavior arising from inter-feature interactions. An efficient geometry based simulation method is used to model the solidification process, and predict global casting soundness. A characterization of the solidification mechanism allows the prediction of centerline macroscopic shrinkage as well as distributed microporosities. The integration of the procedural and rule-based approaches makes this system efficient for predicting castability of parts.

Expert System Applications in
Materials Processing and Manufacturing
Edited by M.Y. Demeri
The Minerals, Metals & Materials Society, 1989

The functions to be performed for the manufacture of castings are (i) Design (ii) Pattern and mold making (iii) Pouring. There is a significant wastage of resources in the design-prototype-redesign cycle between design and manufacturing engineering functions due to lack of techniques to evaluate the castability at the design stage. To minimize the process time and cost, as well as the lead time, an Intelligent Environment called EXCAST has been developed. The purpose of the EXCAST system is to (i) Provide an environment to facilitate the design process, and (ii) Perform Manufacturability analysis on the design to give the designer a feedback about the castability of the design.

Some research has been done in the application of artificial intelligence to the manufacturability of a design. Since most manufacturability knowledge is available in terms of geometrical features, it is important to have feature level geometric information present to efficiently perform the analysis. A significant amount of work has been done in the area of extracting manufacturing features from a 3-D geometric description (database). [1, 2, 3]. However, complete feature extraction by these methods is cumbersome, and unrealistic to be applied in a design scheme. The approach of "Designing with Features" has shown considerable promise as an efficient technique [4, 5]. In this technique, the features that are significant from the manufacturing viewpoint are used as primitives at the design stage. In this work, complex axisymmetric parts are modelled using "Fixed Features" [6] so that (i) Complex geometries can be represented as a group of simplified features, facilitating easy rule based analysis, (ii) A simple and dynamic database is produced for use in other systems, like process planning.

The manufacturability analysis is performed as an integration of knowledge based and procedural methods. Strictly knowledge based systems have been previously developed [5]. Such methods are capable of addressing various details like Parting line generation, Tolerancing, Section thickness checks etc. However, these systems are not capable of addressing the global casting properties, since they cannot account for complicated inter-feature interactions. Currently available knowledge of the casting process is adequate to predict solidification defects from geometric data. However, this knowledge is not in a form that is suitable to be incorporated in a rule base. Therefore, a vital link in the design environment seems to be a model of the process that is used to infer the solidification based defects. Considerable amount of work has been done in the application of Finite Element Methods (FEM) in Casting solidification simulation.[7]. However, the excessive computation burden of this method rules out its use in the current scheme. In this work, an efficient geometry based solidification simulation method is presented. It is coupled with a knowledge based system to accurately predict local and global casting properties. Some work has been done in the area of geometry based simulation [8, 9]. These works are strongly based on the solidification modulus approach, and do not provide accurate estimates of solidification time that could be compared against FEM simulations. Alternative approaches based on the Monte Carlo method have also been proposed [10].

In this paper, an overview of the EXCAST environment is presented. An automotive flywheel is used as a group of parts to illustrate the fixed-features design approach. The fixed features design scheme is described. The knowledge based system built upon this scheme is outlined. The knowledge representation scheme is described, with examples of how this system can detect and predict potential errors in the design. The next section presents an efficient geometric modelling technique and its implementation. This method is presented as an efficient alternative to the FEM technique. The accuracy and time savings of the proposed method are compared against the FEM method. The issue of defect prediction from the results of the simulations are also discussed, and a case study of a flywheel design is presented.

The Fixed Features Design Scheme

In the EXCAST system, axisymmetric parts are grouped into subclasses such as spacers, flanges, flywheels, splines, gears and pulleys. The flywheel group, for example has similar features which are extracted from the cross sectional drawing. These similar features are referred to as "fixed features". Figure 1 shows the features extracted for a flywheel. By extracting these features, it is possible to extract common rules according to the type of features and the topological relationship between them. Fixed feature sets should be the most general feature representation in order to subsume all the members of a part family. By extracting these features, a complicated relationship is simplified, since the topological relationship among the fixed features is predefined. The primary features can also be modified by defining secondary features and cofeatures. Co-features like bosses, holes etc. can be placed on to primary features. Secondary features include fillets for rounding, tapers, etc. located between primary features. The method of extracting these features for a particular subclass is described elsewhere [6, 20].

To create a design using this system, the user first selects the material to be used for the casting and the pattern. The user then proceeds to input the dimensions 1 through 14 from Figure 1. The interactive questioning then continues for the input of secondary and co-features. The geometrical, topological and dimensional description of the part is generated and stored in memory so that EXCAST can retrieve this description into intermediate frames between the memory and the knowledge base. By generating a data file, the designer can view the design using AutoCAD. By using the "Modify" option, the design parameters could be changed. The "Evaluate" option allows the invocation of the expert system using rules for parting line generation, manufacturability evaluation, etc. The system also permits the generation of NC code to drive a 3 axis milling machine for producing a pattern.

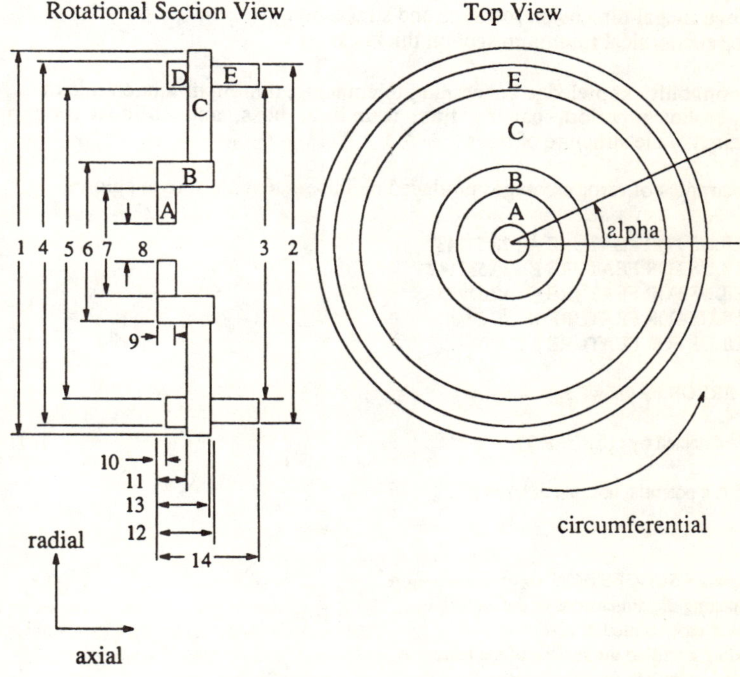

Figure - 1 Extracted fixed-features for a flywheel.

The Knowledge Base

The knowledge base of this system uses a heuristic control strategy to find a path of rules which are invoked sequentially for manufacturability evaluation. The rule firing is guided by gathering rules into sets of meta-rules. Some meta rules used in EXCAST are shown below

```
(META-RULE%CYLINDER (1 2 3 4 5 6))
(META-RULE%CIRCULAR-HOLE
    (15 16 17 18 19 20 21 22 23 24 25 26 27 28 29 30 31 32 33))
```

Manufacturability evaluation for circular hole, for example is carried out using rules 15 through 33. The knowledge in EXCAST is represented using frames and production rules, and is controlled by meta-rules. The heuristic control strategy and the detailed knowledge representation scheme is described in [6, 20]. For instance, rule-50, shown below, checks if a particular section thickness causes manufacturability problems because of fluidity or solidification of molten metal.

```
RULE-50
IF
    Primary feature is 1
  AND
    Direction of hole is left OR right
  AND
    The distance from bottom of hole to side
    is less than minimum economical section thickness
THEN
    Give illegal dimension message and suggestion
    for economical minimum section thickness
```

Manufacturability knowledge concerning tolerancing, minimum section thickness, draft angle, shrinkage, hot junction, parting line, rounding, boss, and solidification modulus are incorporated in the structure of rules.

A few examples of error messages produced by this system are shown below:

```
MODULUS FOR FEATURE A : 28.57143
MODULUS FOR FEATURE B : 20.533142
MODULUS FOR FEATURE C : 40.04975
MODULUS FOR FEATURE D : 20.0
MODULUS FOR FEATURE E : 20.0


--------ERROR.by-knowledge--------

Error is detected by : ($RULE 89)

| There is a potential hot spot at feature A.



    +-----===> SUGGESTION <===---------------+
    | adjusting the dimensions of the feature B
      to increase its modulus.
    | adding a chill to the surface of the feature A
      to lower the effective modulus.
    | attaching a feeder to the feature A.



    --------ERROR.by-knowledge--------

Error is detected by : ($RULE 95)

| Junction between feature C and E are not thermally neutral.
| Possible hot-spot.
```

```
+-----===> SUGGESTION <===---------------+
| Change length of feature E.
| recommended length : less than 150.0
OR
| Add rounding.
| recommended radius : 50.0


--------ERROR.by-knowledge--------

Error is detected by : ($RULE 60)

| Incorrect Parting line for feature C.
| ==> Make combinational parting line with  feature 'C_LEFT'

|-----------------------------------------------------------------------------
| Suggested Parting Line ==> Stepped parting line
| Line 5.0 from (feature C_LEFT) and line 5.0 from (feature A_RIGHT)
|-----------------------------------------------------------------------------
```

EXCAST was written in Common LISP on a Sun Workstation 3/60 UNIX. The system currently contains 152 rules. The system has been interfaced with AutoCAD, and they communicate with each other through a database.

The manufacturability rules used in this work were obtained from interviews with foundry experts, design documents from industry, handbooks, and papers. However, most rules are related to local casting features like rounding, boss, and junction. Global manufacturability can be predicted only qualitatively. In order to deal with global casting properties more accurately, a simulation procedure is discussed in the following sections.

<u>Geometric Simulation</u>

In this section, a method based on time-stepping is described. The effect of the sand mold on the casting is described by using a newly described parameter called the mold cooling factor. It is believed that this method can handle complicated geometries, since it is based on discretization. Also, the element size can be controlled so that the required accuracy is achieved in a minimum of computation time.

<u>Simulation Procedure</u>

The main steps involved in the method are as follows-
1.   Discretize the casting into squares of reasonable size.
2.   Apply the boundary cooling factors.
3.   Select a time step for iterative solution.
4.   Iteratively step through all elements which are still unsolid and update their percentages until they solidify completely. The step at which solidification is complete for each element is noted.
5.   Post process the solidification time data by plotting contours for constant values.
These steps are described in detail in the following sections.

<u>Discretization</u>

The casting is divided into a set of square elements. The discretization procedure consists of breaking up the casting into elements, and initializing the data structures corresponding to the elements. Let the boundary geometry be represented by points (X_1, Y_1), (X_2, Y_2), ...(X_n, Y_n). Let X_{min}, X_{max}, Y_{min}, and Y_{max} represent the coordinate locations of the X minimum limit, X maximum limit, Y minimum limit, and the Y maximum limit respectively of the casting geometry. This is shown schematically in Figure 2. A square of size Max($(X_{max} - X_{min})$, $(Y_{max} - Y_{min})$) whose bottom left corner is located at (X_{min}, Y_{min}) is broken down into square elements based on an element size specified by the user. The elements not present are discarded, and the data structures for the elements present are initialized.

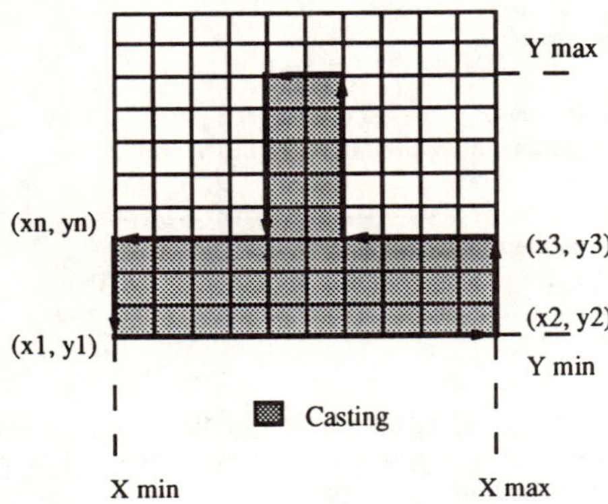

Figure - 2 Discretization scheme for the casting interior.

The neighbor assignment for an element on the boundary is schematically shown in Figure 3. With this procedure, the choice of the element size permits a good approximation to be achieved for castings of complicated geometry. The terminology used for the procedure is as follows:

The <u>Cooling factor</u>, C_e of an element is defined as the effective number of sand elements geometrically identical to it that are available at a particular time, and which determine the cooling rate of an element.

The <u>Percentage Solidifie</u>d, P_e of an element is defined as the ratio of the amount of heat lost by it to the Latent heat.

The <u>Time Solidified</u>, t_e of an element is defined as the ratio of its solidification time index to the solidification time index of the first element to solidify. The solidification time index, i is an integer time step count.

The <u>Casting Solidification Modulus</u>, M is defined as the ratio of the casting volume to the casting cooling area. The Modulus has been used extensively to characterize the solidification behavior of castings in molds.

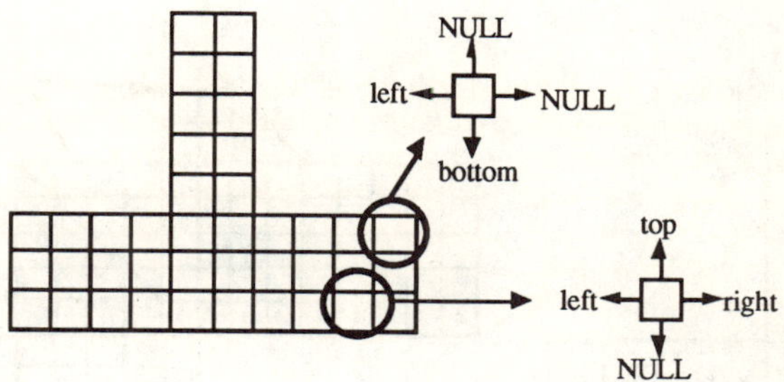

Figure - 3 Neighbor assignment for boundary elements.

<u>Initial and Boundary Conditions</u>

At the beginning of the simulation, the initial conditions corresponding to pouring are to be set in the model. For all the elements present within the casting, the Percentage Solidified, P_e is set equal to zero at the beginning of the simulation. The Cooling factors, C_e which decide the rate of solidification of the elements are also to be assigned to the boundary elements at the start of the simulation.

An assumption is made that a sand mold of thickness equivalent to that necessary to completely absorb the heat content of the casting is present around the casting. The thickness of the mold can be easily calculated from heat balance equations.

The mold is divided using the same element size as is used to discretize the casting. The assumption is that a casting element which is in contact with one sand element, and consequently has a cooling factor of unity, will take one unit of time to solidify completely. This assumption means that the total cooling power of the mold is exactly equal to that necessary to solidify the whole casting completely, because of the heat balance assumption made in calculating the mold thickness.

The cooling factors are initially assigned only to the boundary elements. After the boundary elements solidify, they sequentially propagate the cooling factors to their neighbors that are still not completely solid. The initial cooling factor assignment is made using some rules regarding external and internal corners. Figure 4 shows a scheme to assign cooling factors to elements. In general, the cooling factors for external corners are much higher than the sand thickness, since they are in contact with a large area of the mold, and the heat flow field diverges out into the sand. At internal corners, the cooling factors are small, since the heat flow field is convergent, and the mold is prone to getting quite hot.

<u>The Time Stepping Algorithm</u>

As described earlier, the assignment of the cooling factors to the boundary elements is used to determine the initial cooling rates of the elements. The time stepping algorithm marches through time using a small step, Δt that is decided by the user. An integer time step count, i is maintained for marking the element's solidification time. At each time step, the percentage solidified for each unsolid element is updated by the value $100 \, (C_e \, . \, \Delta t \,)$. If an element crosses over 100% solid fraction, the current integer time step count, i is noted in the element's data structure. The cooling factor of this element is then distributed to the unsolid neighbors, after subtracting a value of 1.0 consumed for its solidification. After all elements have solidified, their time indices are normalized by dividing through into the minimum value, so that a non dimensional time estimate, t_e is obtained.

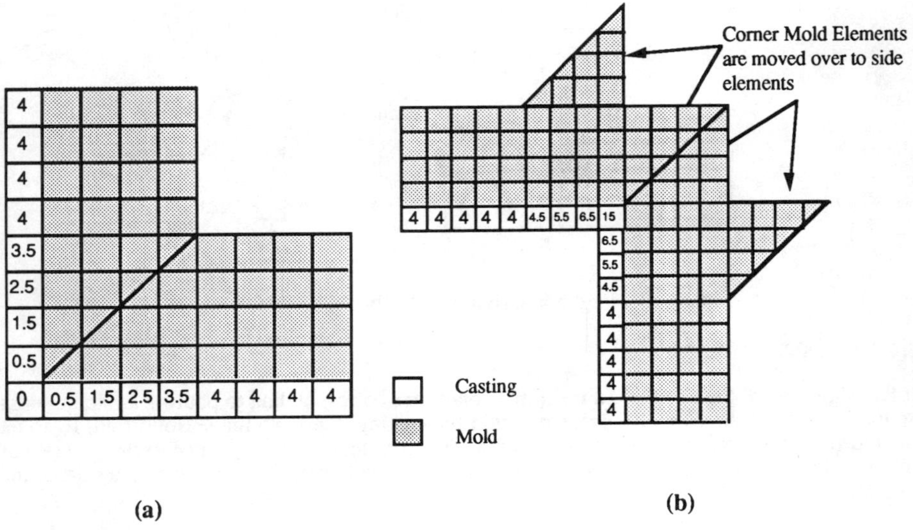

(a)   **(b)**

Figure - 4 Cooling Factor Assignment for boundary elements. (a) Internal
Corner   (b) External Corner.

<u>The Flat Plate Analogy</u>

After time stepping, we are left with non-dimensional time count , t_e for each element, which
goes from 1.0 for the first solidified element to t_{max} for the element to solidify last. Conversion
of the time count to real time units is done by comparing the shaped casting to a flat plate of
equivalent modulus. It has been shown, for the case of short freezing range alloys, [11] that the
Modulus is directly related to the solidification time. Therefore, a flat plate of equivalent modulus
should have the same solidification time as the shaped casting. According to the Chvorinov rule,
the solidification time, t is dependant upon the modulus, M as

$$M \text{ (inches)} = k \ \sqrt{t} \text{ (sec)} \tag{1}$$

The constant k is independent of geometry for the case of narrow freezing range alloys. The
value of k is obtained using the analytical solution for the case of a pure metal solidifying in a
mold. This analytical solution is available for the case of a semi infinite plate of metal in a semi
infinite mold [12]. The solution is obtained by solving an error function equation.

Let the parameter t_e have values ranging from 1.0 to t_{max}. The element's t_e will then correspond
to a flat plate Modulus of (t_e / t_{max}) M, where M is the Modulus of the shaped casting. This is
because the index i at the time of simulation corresponds to a factor that is linearly related to the
Modulus value (Figure 5). The equivalent times of solidification are calculated by marking off on
the solidification plot of the flat plate of the Modulus specified above. In this manner, the
parameter determined from the simulation, t_e can be directly converted to solidification time using
the flat plate analogy. In practice, it may be difficult to obtain a value of equivalent modulus for a
shaped casting. If an accurate estimate of solidification time is available from FEM or any such
procedure, an equivalent modulus can be calculated. Otherwise, the value of the Volume/Cooling
Area of the casting can be used as a rough estimate.

Figure 6 shows a comparison between the end of freeze isochrones of Geometry and FEM based
simulations for an L section. The metal was assumed to have a freezing range of 10°C.

<u>Dependance on Freezing Range</u>

The comparison between geometry and FEM based simulations for the end of freeze isochrones is accurate only for narrow freezing alloys. Chvorinov's rule seems to hold only for pure

metals, and the solidification mechanism for long freezing range metals exhibits significantly different behavior. The solidification times predicted by the flat plate analogy for long freezing alloys shows a considerable dependance on geometry. This is contrary to Chvorinov's rule, which states that Solidification time depends on the Modulus, regardless of the shape.

FEM simulations of flat plate solidification for 0.38% C Steel were compared against reported experimental results on the solidification of a 7" square cross section [13]. The Solidification histories are compared in Figure 7 and are seen to be considerably different. As a check, FEM simulation of the reported square casting was seen to give a result that was acceptably close to the reported experimental data. Flat plates of different thickness were simulated, and these resulted in different Modulus vs. Time histories. Some work has been reported [14] in which a polynomial regression fit of the solidification of a 7" square was taken as a basis to determine the Modulus vs. Time dependance for wide freezing alloys. The results of FEM simulations of the current work seem to suggest that the solidification times cannot be characterized uniquely based on Modulus alone, and the geometry of the casting also plays an important role.

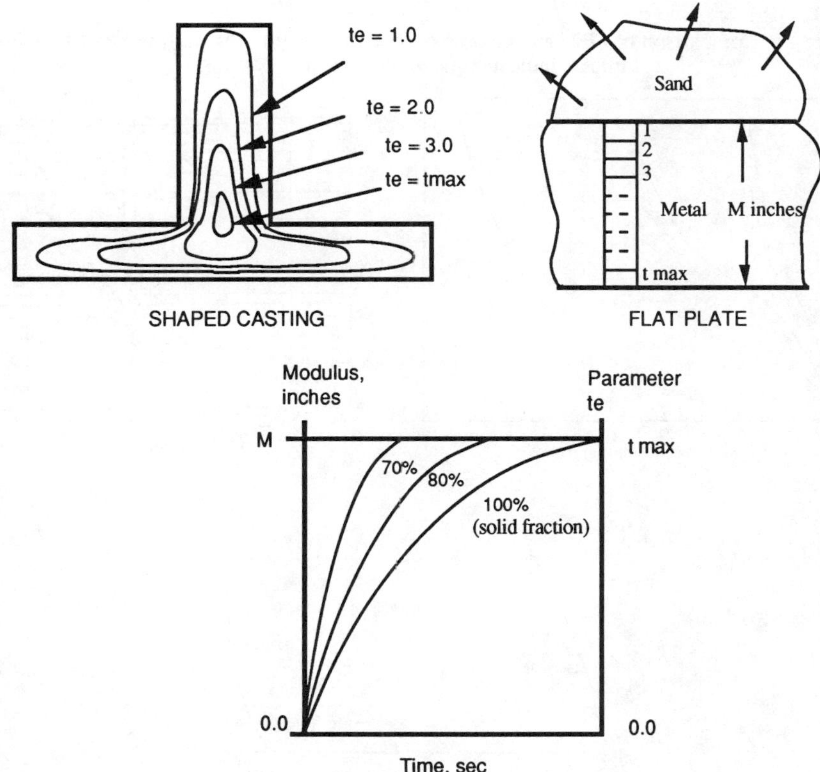

Figure - 5 Analogy Comparing the Shaped Casting against the Analytical Solution for Flat Plate Solidification. This is used to determine the actual solidification time for the Casting.

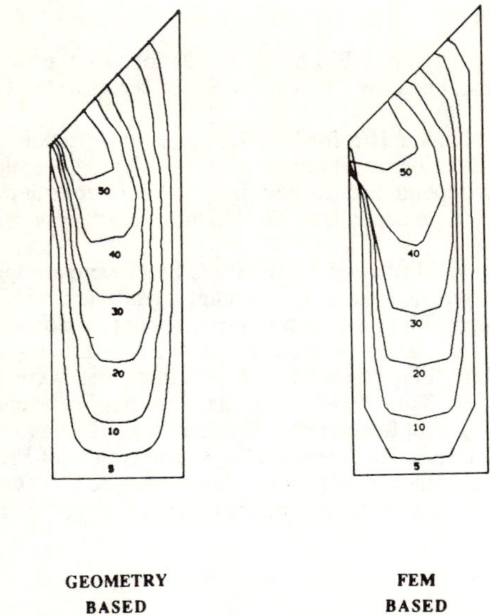

Figure - 6 Comparison of FEM and Geometry based simulation for Narrow Freezing Alloy. Numbers indicate time in minutes after pouring.

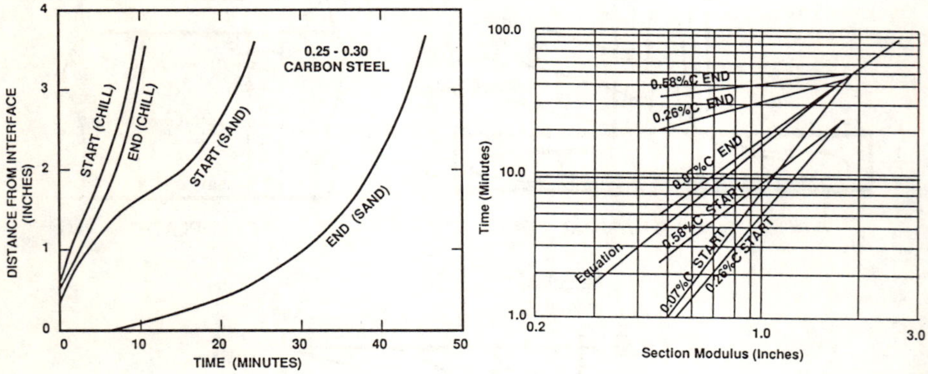

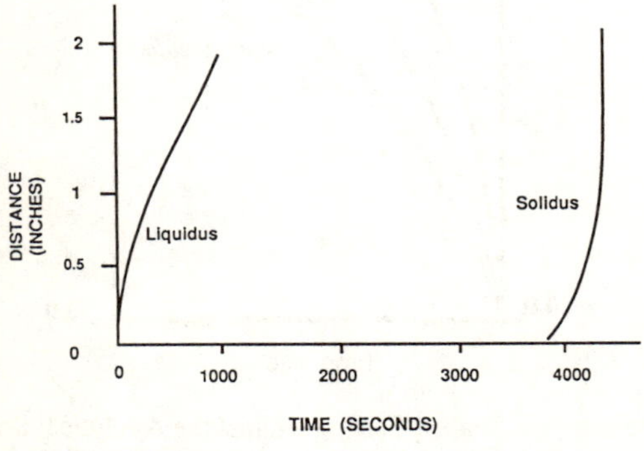

Figure - 7 Limitations of Modulus based Techniques in simulating wide-freezing alloys.

However, as will be later discussed, in the case of wide freezing metals, the critical solid fraction that is used to predict defects is about 70%. For this value of f_{cr} , the metal can be assumed to have nearly narrow freezing behavior. Figure 8 shows the comparison for this case, where the freezing range is 1451-1508 °C. Thereby, the accuracy of this method for predicting the movement of critical solid fraction contours is verified.

As a test case, the simulation was carried out for a flywheel design. The results are shown in Figure 9. The metal is 0.38 % C Steel with a wide freezing range of 1451-1508 °C. The analytical solution for the flat plate analogy assumes an isothermal freezing metal with the Latent heat corresponding to 0% - 70% solid fraction enthalpy difference.

<u>Defect Prediction</u>

When feeding metals and alloys, it is important to consider the mode of solidification. The extremes of solidification behavior are plane front solidification or skin freezing, that occurs in the case of pure metals or narrow freezing alloys and mushy freezing behavior that is exhibited by long freezing range alloys.

In the case of pure metals, growth takes place from the mold wall with a plane solidification front. Late in the freezing period, a channel or pipe will be formed in the middle of the casting. This type of piping defects occur along the geometric center of the casting, and are referred to as *center line shrinkage*. Minimal shrinkage defects are observed away from the center line in the case of narrow freezing metals.

In the case of mushy freezing alloys, the temperature gradients in the casting are very small. There will be only very small variation in temperature or solid fraction across the whole casting. The entire casting is mainly a uniform *mush*. In this case, the distance to be fed to reach the portions that are solidifying is quite large, and a large pressure drop occurs before liquid metal can be fed to compensate for shrinkage. The nature of defects in this case is *distributed microporosities* centered at the hot spots in the casting.

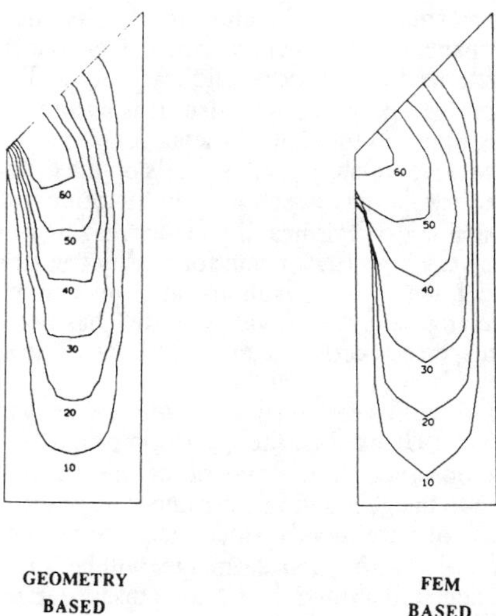

Figure - 8 Comparison of FEM and Geometry Based Simulation for a wide freezing alloy. Critical solid fraction of 70% is used for the comparison.

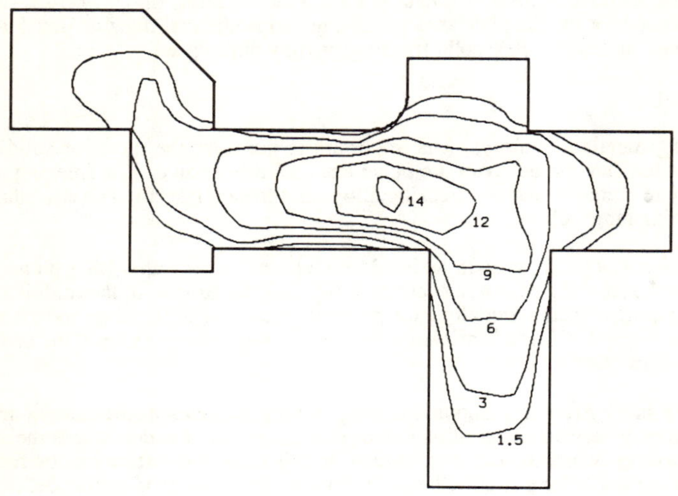

Figure - 9 Result of Geometry based simulation of a Low Carbon Steel Flywheel. The freezing range of the material is 57°C. Numbers show time in minutes after pouring.

Figure 10 shows different attempts in literature to classify metals based upon solidification behavior. There are discrepancies with regard to in which class different alloys should be placed. Also, the alloys are characterized by a critical solid fraction f_{cr}. This critical value is identified as the fraction solid at which feeding problems arise. It is assumed that for solid fractions lower than critical, there is full fluidity in the molten metal, for values higher than critical, the regions are prone to shrinkage. The level of shrinkage depends on the distance over which feeding has to take place. Various values for f_{cr} have been used in literature, ranging from 70% to 95% (eg. [15], [16]). The critical fraction depends heavily on the freezing mode. It is clear that the value to be used for pure metals is very high (99%), and for mushy freezing alloys is very low (70%). In the case of low carbon steel, which is the main metal of interest in this study, literature seems to indicate intermediate freezing behavior. A value of 70% has been assumed in lot of work. The main type of defects are dispersed porosities, although center line shrinkages are not uncommon.

The freezing mechanism identifies the relative importance between microshrinkage and center line macroshrinkage defects. Using this, the f_{cr} loop progress will help to localize the areas where the defects will be observed. In the case of center line macroshrinkage, they should be localized in the areas close to the geometric center line, and getting more severe with the progress of the f_{cr} loop. In the case of pure metals, since there are no feeding problems, a central pipe whose volume is equal to the solidification shrinkage will be formed at the end of solidification. Distributed porosities are centered around the f_{cr} loop disappearance point. [15]. For quantitative prediction of the extent of distributed porosities, feeding range calculations are to be performed to decide the extent to which the solidifying metal can be fed to soundness. Such calculations are available only for simple shapes like a cylinder with end feeder, eg [16, 17, 18, 19]. Such calculations for shaped castings are beyond the scope of the present implementation. Therefore, in the case of wide freezing alloys, the region between solid fraction of f_{cr} and 100% is identified as a possible region for the occurrence of distributed microporosities.

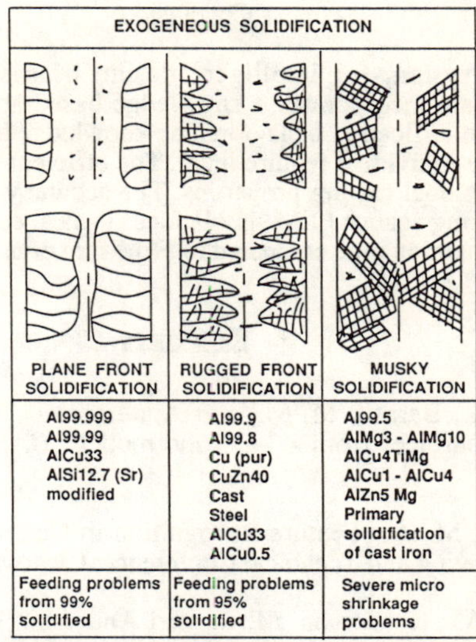

Figure - 10 Classification of Metals based on Solidification behavior, from Davies [16].

Figure 11 shows the defect prediction for a steel flywheel using the current method. The regions in which macroscopic as well as microscopic shrinkages could occur have been identified. The relative importance of macro and micro shrinkage can be estimated from the freezing range of the material.

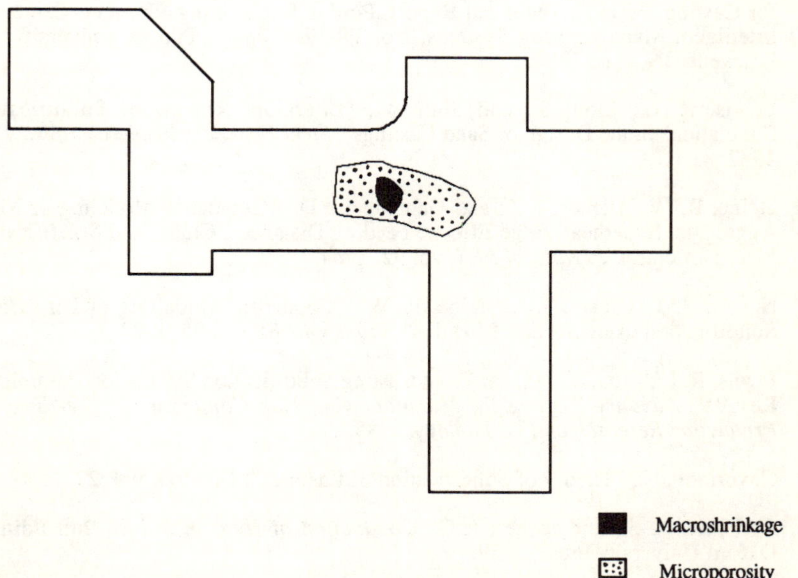

Figure - 11 Defect Prediction using the Results of Geometry based Simulation of Flywheel Solidification.

Conclusion

An integrated system for intelligent design of castings has been proposed and implemented. The system incorporates a knowledge based approach as well as a procedural technique to simulate solidification behavior. The knowledge based system addresses the local casting properties at the individual feature level. The efficient simulation technique developed here is used to predict global casting properties. The accuracy of the simulation technique has been verified by comparing against FEM simulations. Typical casting geometries resulted in over 10 fold savings in computation time compared to FEM for comparable accuracy.

References

1. Choi, B. K., Barash, M. M., and Anderson, D. C., "Automatic Recognition of Machined surfaces from a 3-D solid model", *CAD*, v16, no. 2, pp 81-86, March, 1984.

2. Henderson, M. R., "Feature Recognition in Geometric Modeling", CAM-I's 13th Annual Meeting and Technical Conference, Clearwater, Florida, Nov 13-15, 1984.

3. Staley, S. M., Henderson, M. R., and Anderson, D. C., "Using Syntactic Pattern Recognition to Extract Feature Information from a Solid Geometric Database", *Computers in Mechanical Engineering*, pp 61-66, September 1983.

4. Libardi, E. C., Dixon, J. R., and Simmons, M. K., "Designing with Features: Design and Analysis of Extrusions as an Example", Proceedings, *Mechanical Engineering Design Conference*, Chicago, IL, March 1986

5. Luby, S. C., Dixon, J. R., and Simmons, M. K., "Creating and Using a Features Data Base", *Computers in Mechanical Engineering*, pp 25-33, Nov. 1986.

6. Chu, C. N., Kashyap, R. L., and You, I. C., "Knowledge Based Expert System for Casting Design", Technical Report, Purdue Engineering Research Center for Intelligent Manufacturing Systems, No. TR-ERC 88-13, Purdue University, W. Lafayette, IN, June 1988.

7. Lewis, R. L., Liou, S., and Shin, Y., "Literature Review of Solidification Simulations in the Design of Sand Castings", *Steel Founders Research Jour.*, v.17, 1987.

8. Heine, R. W., Uicker, J. J., and Gantenbein, D., "Geometric Modeling of Mold Aggregate, Superheat, Edge Effects, Feeding Distances, Chills, and Solidification Macrostructures", *Trans. of AFS*, vol 92, 1984.

9. Neises, S. J., Uicker, J. J., Heine, R. W., "Geometric Modelling of Directional Solidification using Section Modulus", *Trans. of AFS*, vol 95, 1987.

10. Lewis, R. L., Liou, S., "Planar Front Casting Solidification Simulation Preliminary Results", *Advance Systems for Manufacturing 12th Conference Proceedings on Production Research and Technology*, 1985.

11. Chvorinov, N., "Theory of Solidification of Castings", *Gisserei*, vol. 27, 1940.

12. Carlslaw, H. S., and Jaeger, J. C., *Conduction of Heat in Solids*, 2nd Edition, Oxford University Press, 1959.

13. Brandt, F. A., Bishop, H. F., and Pellini, W. S., "Solidification of various Metals in Sand and Chill Molds", *Trans. of AFS*, v. 62, pp 646-653, 1954.

14. Dekalb, S. W., Heine, R. W., and Uicker, J. J., "Geometric Modeling of Progressive Solidification and Cast Alloy Macrostructure", *Trans. of AFS*, vol. 95, 1987.

15. Imafuku, I., and Chijiiwa, K., "A Mathematical Model for Shrinkage Cavity Prediction in Steel Castings", *Trans. AFS,* vol. 91, 1983.

16. V. de L. Davies, "Feeding Range Determination by Numerically Computed Heat Distribution", *AFS Cast Metals Jour.*, June, 1975.

17. Campbell, J., "Feeding Mechanisms in Castings", *AFS Cast Metals Res. Jour.*, v. 5, no. 1, pp 1-8., March 1969.

18. Piwonka, T. S., Flemings, M. C., "Pore Formation in Solidification", *Tr. Met. Soc. AIME*, v. 236, August, 1966.

19. Walther, W. D., Adams, C. M., and Taylor, H. F., *Trans. of AFS*, v. 64, p 658, 1956.

20. You, I. C., Chu, C. N., and Kashyap, R. L., "Expert System for Castability Evaluation: Using a Fixed-Features based Design Approach", *Robotics and Computer Integrated Manufacturing*, v. 6, 1989.

DESIGNING AN EXPERT SYSTEM FOR THE PRODUCTION OF

SILICON CARBIDE WHISKERS

W. J. Parkinson, P. D. Shalek, and E. J. Peterson

Los Alamos National Laboratory
P. O. Box 1663
Los Alamos, New Mexico 87545

G. F. Luger

Department of Computer Science
University of New Mexico
Albuquerque, New Mexico 87131

Abstract

Silicon carbide whiskers, a very strong material produced primarily for strengthening ceramics and metals, are being considered for various commercial uses. Whisker production is a semi-batch process that is difficult to model mathematically, so rules accumulated from experience are used to set up and run the process. Several sets of process conditions are possible. When the correct set is chosen, the proper whisker for the desired use is produced. We designed a two-phase, PC-based expert system to help inexperienced users. In Phase I, an expert consultant provides users with information that enables them to set up the run. This information is incorporated into the rule base, which makes up the second phase, the control system. Because this project is laboratory scale, no automatic controls were added to our system. Instead, we use a human controller: the operator. The operator asks the expert system if the process is behaving correctly. The expert system makes the decision and suggests any corrections the operator should make. With this system design, the automatic controls can be added later.

Expert System Applications in
Materials Processing and Manufacturing
Edited by M.Y. Demeri
The Minerals, Metals & Materials Society, 1989

Introduction

Silicon carbide whiskers are a very strong material that resemble cat's whisk-ers. They are produced primarily as a reinforcing material for strengthening ceramic or metallic composites. Whiskers can be used as randomly oriented chopped fibers, or they can be grown in long lengths, which can be made into yarns and woven. When woven, these fibers create an even more effective directional reinforcement. Although the primary purpose of the whiskers is for compositing materials for strength, other uses are also being considered.

Whisker production is a semi-batch process that is extremely difficult to model mathematically. We are using rules accumulated from many years of trial and error experience to successfully set up and run the process. Two catalyst types, two reactor configurations, and several sets of process conditions are available. When the proper combination of these variables is chosen, the desired result can be attained.

In the past, as long as one person would set up and operate the whisker production runs each time, it was considered adequate for that person to keep the rules in his head. But because the whiskers process is now a candidate for technology transfer, we are faced with the problem of how to transfer the expertise to industry without transferring the expert. We therefore designed an expert system to organize and assist in the solution process.

The first phase of the expert system design was to build an expert consultant to provide the user with enough information to correctly set up the run and produce the desired results. The set-up information was then incorporated into the rule base to make up the second phase, the control system. The control system corrects perturbations in the process conditions to ensure that the proper corrections can be made and to ensure that the desired product will be obtained at the end of the run.

Because our system is a "laboratory-scale" project, we have not added elaborate sensory devices to test for perturbations or upsets. Instead, we use a human sensor: the operator. After observing the controlled variable read-ings, the operator asks the expert system whether the process is behaving correctly. The expert system responds and suggests which, if any, corrections should be made. The operator then makes the corrections. This system is designed so that automatic controls and sensors can easily be added at a later date or for a larger operation.

Most of the rules for the expert system were obtained by talking to the operators, and some (especially those that involved product quantity and quality) were obtained from a well-designed relational database. This data-base allows us to observe and plot data in a wide variety of patterns. Using classification methods from pattern recognition theory, we were able to find the optimum correlations, and therefore, to design the best rules.

The expert system is rule-based, and it is designed to run on a PC. For easy access by the operator, the PC is kept in the laboratory associated with the whisker process equipment.

<u>**The Silicon Carbide Whisker Growth Process**</u>

On the laboratory scale, the silicon carbide whisker process is a semi-batch
process. That is, part of the process is run in the transient batch mode and
part of the process is run in the steady-state continuous flow mode. Figure 1 is
a diagram of the silicon carbide whisker reactor. Silicon dioxide bricks
impregnated with graphite are placed inside the reactor. After they are
heated, these bricks produce silicon monoxide by reaction (1).

$$SiO_2 + C \rightarrow SiO + CO \ . \tag{1}$$

The SiO (the ingredient used to make silicon carbide) production rate is proportional to the concentration of SiO_2 and CO in the brick, and those concen-
trations are diminishing with time. This is the transient batch portion of the
process. As shown in Fig. 1, a mixture of gases containing methane, the carbon source for the silicon carbide production, is forced through the reactor.
For many situations, the composition and flow rate of this gas mixture does
not vary throughout the length of a production run. This is the steady-state
portion of the process. The silicon carbide is formed by reaction (2).

$$SiO + CH_4 \rightarrow SiC + H_2O + H_2 \ . \tag{2}$$

An idealized growth sequence for a silicon carbide whisker is shown in Fig. 2.

The SiO formed by Reaction 1 must mix with CH_4 in the process gas stream.
These gases must find their way to a whisker growth surface, as shown in
Fig. 2. Flow-visualization studies (1) have shown that this path is tortuous,
involving several different modes of mass transfer. Figure 3 shows the modes
of mass transport from the SiO generator to the whisker growth surface.
Figure 4 shows the steps involved in the overall kinetic process for growing the
whiskers.

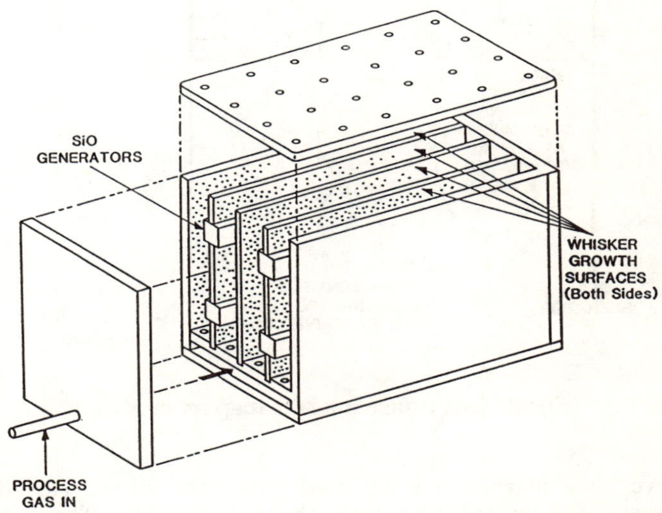

Fig. 1. The Los Alamos silicon carbide whisker production reactor.

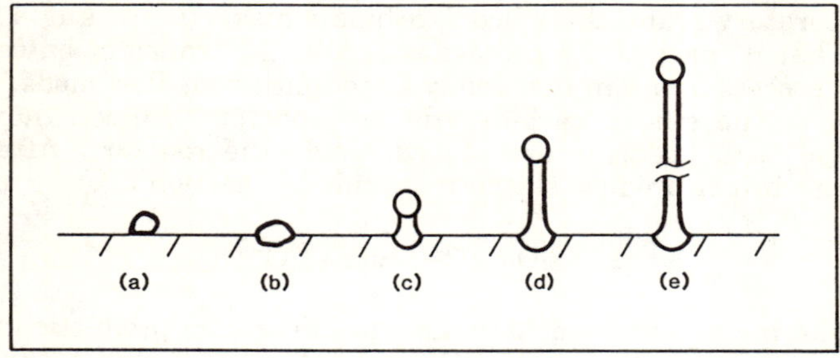

Fig. 2. Idealized growth sequence for the Los Alamos silicon carbide whisker production: (a) metallic catalyst is melted on a substrate, (b) catalyst forms a crater in the substrate, (c) silicon carbide whisker is nucleated, and (d,e) whisker grows away from the substrate with liquid catalyst ball at the tip.

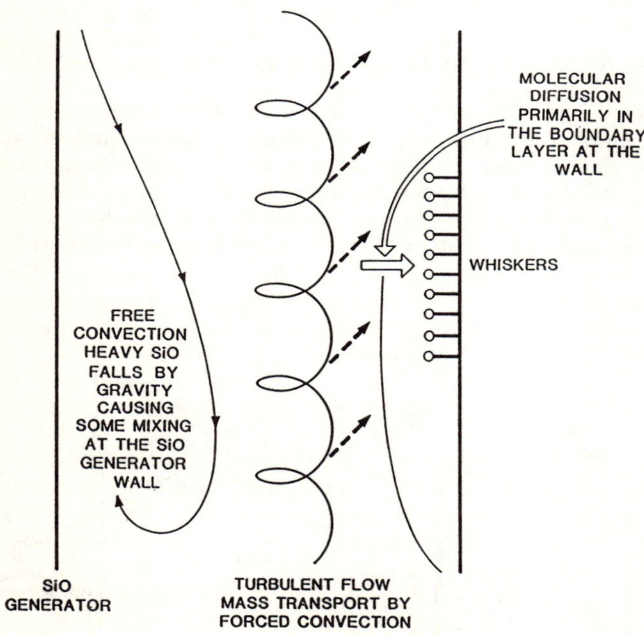

Fig. 3. Gas-phase mass transport modes.

Although we have learned a great deal about the silicon carbide whisker growth process and we can now grow them quite well. The foregoing discussion indicates how hard this process is to model with normal mathematical-algorithmic techniques. In fact, we first tried the mathematical

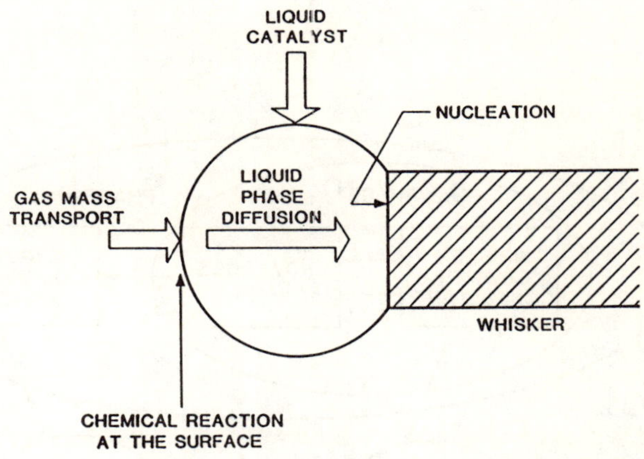

Fig. 4. Steps in the overall kinetic process for growing silicon carbide whiskers.

modeling approach and found that our models would not adequately predict whisker yield and type from a particular experimental setup. Furthermore, they were not adequate for process control. We had to ask the question "How can we grow whiskers as well as we do when we don't understand the physics and chemistry of the process any better than we do?"

The answer to this question is that our expert operators have learned excellent rules of thumb, through years of trial-and-error experiences, for setting up and running whisker growth experiments. Our next question was "How do we capture this expertise so that the process can be set up and run by people who are not experienced experts?" This question is especially important since the technology of the whiskers process has been earmarked for transfer to industry. We would like to be able to transfer the technology without transferring our experts as well. So the answer to this question was to write an expert system or systems to capture this expertise.

The Expert Systems

Our goal for this stage of the project was to develop a small PC-based expert system in two phases, as shown in Fig. 5. In Phase I, we developed a whisker growth consultant to help the operator set up a whisker run for the desired whisker type and quantity. The knowledge developed during this phase was entered into a knowledge base that could be used with the expert control system. The expert consultant developed in this phase can be used with the expert control system developed in Phase II, or it can stand alone. The expert control system developed in Phase II uses rules to control our laboratory-scale silicon carbide whisker growth process. The expert control system is somewhat limited because it can not work without an operator interface. This scenario is depicted in Fig. 6. Although this system is adequate for our current process, it would fall short for a full-scale whisker production plant. For this reason, we did some exploratory work with a system that may work for a full-scale plant. This scenario is depicted in Fig. 7. Even though the work was brief, the results were very interesting. They are discussed at the end of this section.

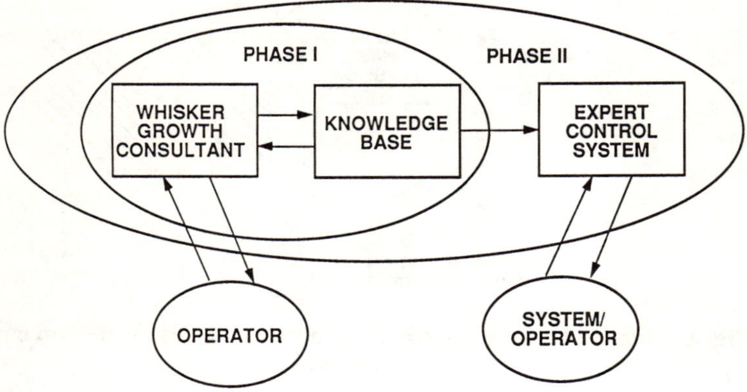

Fig. 5. The two phases of the expert system design.

Fig. 6. Demonstration of the operator interface between the current expert
system and the process.

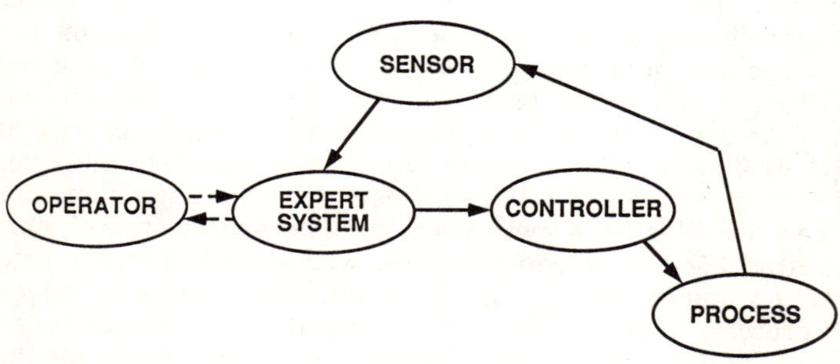

Fig. 7. Desired expert system/process interface for full-scale plants.

We felt that, in this late 1980s time frame, any usable expert system should be executable with an inexpensive shell and an easily available computer. We therefore focused our search on rule-based shells for the PC family of computers. But we were not sure that the available PC-based shells would be adequate for the expert system we envisioned. We were pleasantly surprised when we found two shells that could handle the task at an affordable price.

After our extensive search for an appropriate PC-based shell, we made two observations. The first was that powerful expert system shells are expensive, and when written for use on a PC, they often include special purpose computer chips that add special features to the PC. The costs of these shells varied from just under $1000 to several thousands of dollars. The second observation was that PC-based shells that cost around $100 were not adequate for our job. We were fortunate to find two shells, CLIPS (2) and EXSHELL (3) that met our low-cost criteria and were adequate for handling our problems. Both shells were supplied to us essentially without cost, and both have performed well. Neither has shown any sign of degradation or reduced performance with our applications.

CLIPS was developed by NASA. It is a forward chaining, rule-based shell written in the C programming language. To use the CLIPS shell it is helpful, though not essential, to know both the C and LISP programming languages.

EXSHELL was developed by the University of New Mexico Computer Science Department. It is a backward chaining, rule-based shell written in the PROLOG programming language. One must know some PROLOG to use EXSHELL.

The shells are different. Both have strengths and weaknesses, but both are adequate for this job. The work shown here is from the CLIPS shell simply because we have worked more with CLIPS at this point.

Figure 8, taken from Ref. (4) shows an empirical phase diagram for the growth, and for the types of whiskers that can be produced as a function of gas composition. The abscissa represents a change from silicon-rich to carbon-rich gas mixtures. The ordinate represents the silicon monoxide concentration in the gas phase. The properties of the whiskers from categories one through seven, shown at the top of the chart, depend primarily upon the whisker diameter. To date, there is some commercial interest in all sections of the chart except areas E and F, the combined species and the large bent needles.

For this study, we have lumped the whisker types into slightly different groups based on lengths and diameters. The whisker lengths vary from about 1/8 in. to about 3-1/2 in. We have divided them into three categories: short, medium length, and very long. The whisker diameters vary from 0.1 to 15 μm. We divided them into three groups: small, medium, and large. With short whiskers, we are interested only in small diameters. With long and medium length whiskers, we are interested only in those with medium and large diameters.

In addition to developing rules to produce a particular whisker type, we have developed some rules to help maximize the production of those whiskers

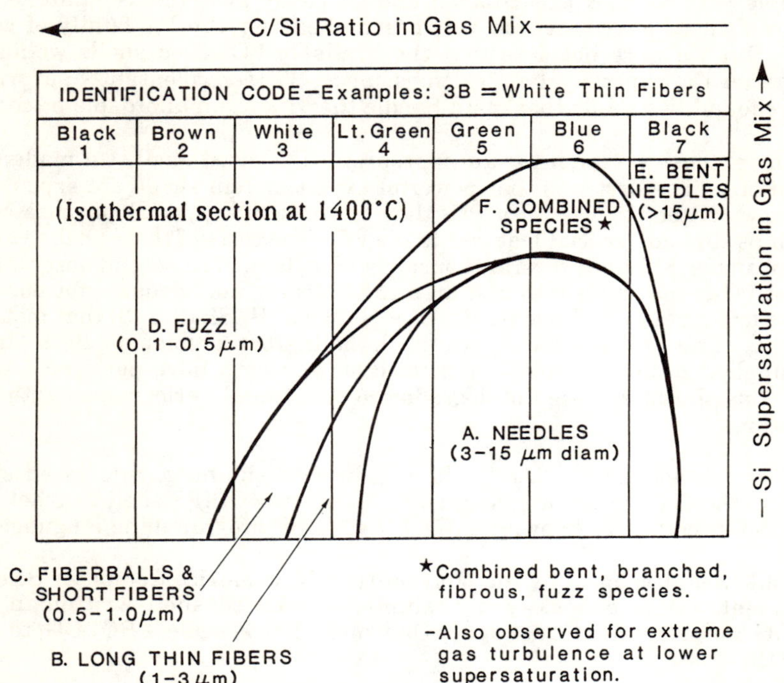

Fig. 8. Empirical phase diagram for the growth of silicon carbide whiskers.

under various operating constraints. These rules can be divided into three categories: (1) how to obtain the maximum yield with new growth plates, (2) how to obtain the maximum yield with used growth plates, and (3) how to obtain a maximum yield in a limited run time. We can also have combinations of Category 1 or 2 with Category 3. These rules depend upon the type of whisker that we are trying to produce, and they are included in our whisker growth consultant expert system.

For our laboratory scale operation, we have been primarily concerned with Categories (1) and (2). Different gas compositions are required with new growth plates than with used growth plates to produce the same quantity of whiskers. After one run, the new growth plates are coated with silicon carbide, and from then on, the silicon carbide participates in the process chemistry. After about four runs, the whisker production degrades to the degree that we must replace the plates. We have developed some cost analyses of our process and find it to be labor and material intensive. Replacing growth plates after every run is too expensive, even for a "laboratory-scale" process. Figure 9 shows the general shape of the time-vs-whisker yield curve. Because we are a laboratory operation, our approach is to run the reactor as long as possible to produce the maximum yield.

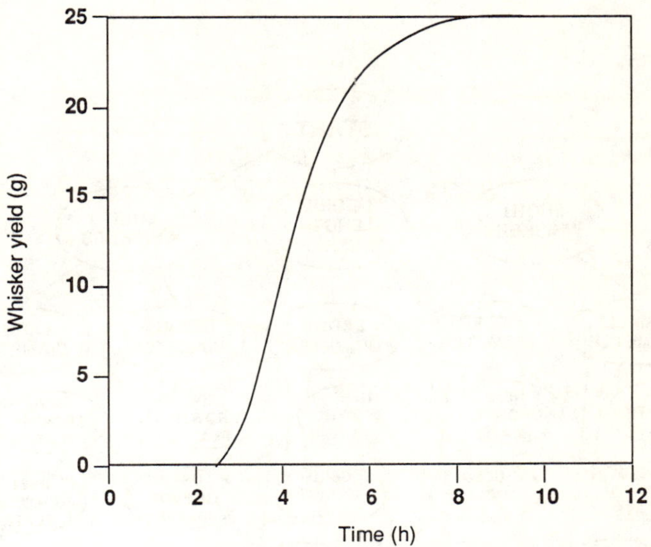

Fig. 9. A time-vs-yield curve for the silicon carbide whisker growth process.

Figure 9 suggests that a point of diminishing returns is reached before the reactor is shut down. Because this process is a candidate for technology transfer to industry, we have developed some production rules for maximizing yields with shorter run times based on the curve shown in Fig. 9. We assume that an industrial process, even if it is a batch process similar to ours, would be optimized in a different manner. For example, if several batches were run in one day, to the point of diminishing returns, more whiskers would be produced than in our previous maximum production run, yet in the same amount of time.

Figure 10 is a simplified search tree for our whisker growth consultant. The leaves of the tree represent operating conditions that will produce the types of whiskers we want to make. We can change the production from medium length to long whiskers by changing the reactor and catalyst type. To produce short whiskers, we must change the gas composition. Whisker diameter depends primarily on catalyst choice and particle size. At first glance, picking the proper production rules for a given run seems straightforward. However, rules obtained from our database have shown that this procedure is more complicated to set up and run optimally than it appears at first observation. For example, gas compositions and temperatures should be different, depending upon the catalyst and the particle size used. The rules for maximizing the whisker yield in a shorter period of time depend upon the intended whisker diameter and length. Long whiskers are not normally produced in shorter run times, and so on.

For the expert control system to work, it must have the information supplied to the knowledge base by the operator and the expert consultant. Because the sensors in our system do not communicate directly with the expert control system, the operator must observe the sensor output, communicate with the control program, and then, if necessary, manually adjust the system controls.

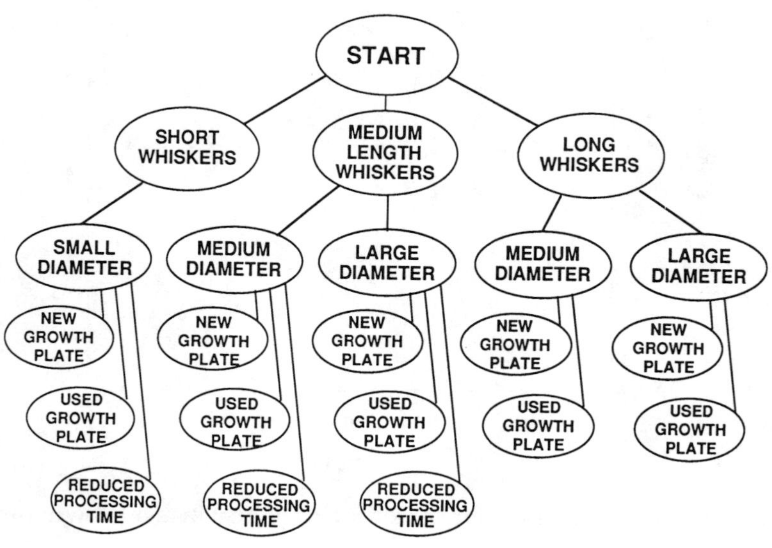

Fig. 10. A search tree for the whisker growth consultant.

The current system has only eleven sensed variables and eight possible control adjustment actions. Three of the sensed variables are temperature, pressure, and total inlet flow. The other eight variables are the inlet and outlet compositions of the four process gases, hydrogen, carbon monoxide, nitrogen, and methane. The eight possible control adjustment actions are as follows: shutdown; adjust temperature; adjust total inlet flow; adjust the flow of the individual inlet gases, hydrogen, carbon monoxide, nitrogen, or methane; or take no action. The amount of adjustment is highly dependent upon the current reading and the run set-up conditions supplied by the expert whisker growth consultant. Figure 11 shows the search tree for the expert control system.

Figure 12 is a simplified search space diagram for our expert consultant. The rectangular blocks represent the major decision points in the program. Figure 13 shows the CLIPS version of a dialogue with the whisker growth consultant. It follows the search space shown in Fig. 12. Some of the values given in Fig. 13 are fictitious because the whisker process falls under the purview of the United States Export Control Laws and some information is subject to limited access. The questions in Fig. 13 are questions asked by the expert system. The answers to the questions are user supplied. Note that, at each decision point, the user is given the opportunity to use values other than those recommended. The system was set up this way because, at the end of the consulting session, the correct values must be available for use by the control system.

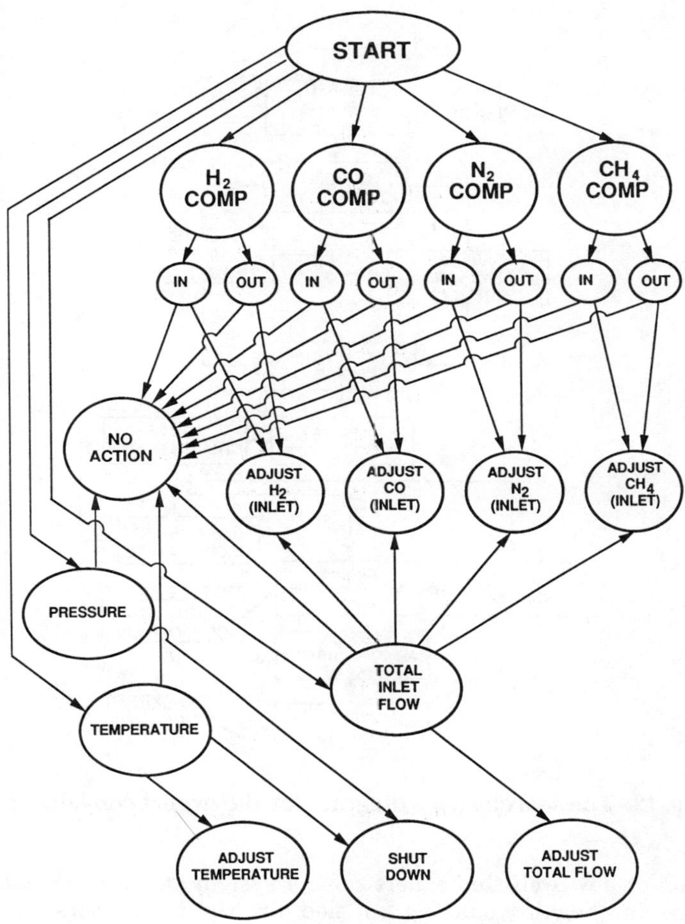

Fig. 11. A search tree for the expert control system.

Figure 14 is a search space-diagram for the expert control system. Figure 15 shows the CLIPS version of a dialogue with the expert control system. Note that the control system makes many of its decisions based on parameter values given during the session with the expert consultant.

It should be pointed out that our laboratory scale process is relatively simple. Our operation is not subject to some of the problems of a full-scale plant, such as real-time time-constraint problems caused by the need to search through thousands of rules before an urgent decision can be made. Our expert system works fast enough for the whisker process at this scale. Evidence for this is given in the last line of Fig. 15: "Reduce the inlet composition . . . and check with me again in 40 minutes." In the laboratory environment, gas-stream compositions are only monitored every 40 minutes. This is adequate.

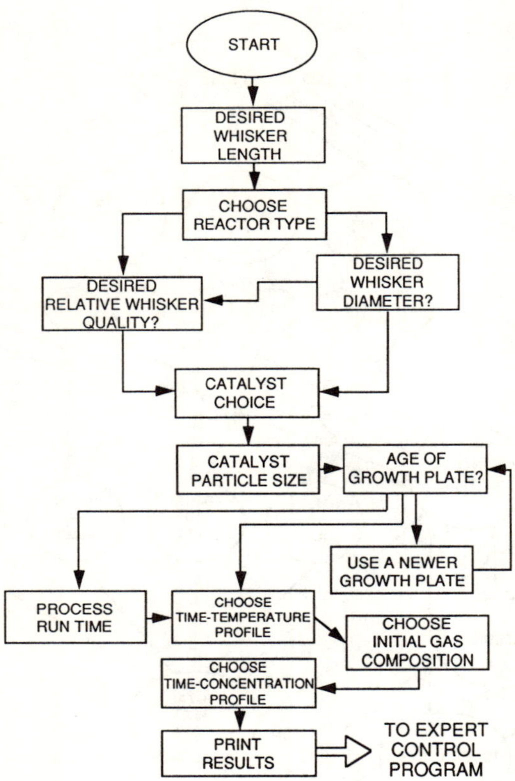

Fig. 12. The search space diagram for the expert consultant.

The question of how well this expert control system would work with a full-scale process and how it would work if tied directly into sensors and controllers is a difficult one. Even though it was not in the work plan for this project, we looked briefly at this question by writing our control system in a simple object-oriented language called OOPS (3). OOPS is written in the LISP programming language and stands for Object Oriented Programming System. Our system was simulated as shown in Fig. 7. The sensors, controllers, and process were simulated as closely as possible by the program. The expert system rules were built into the program. A diagram of the model is shown in Fig. 16. The program was run several times and responded to simulated sensor inputs with correct control to the simulated process. This experiment verifies that our expert control system does not really require an operator interface. It was exciting to observe how well an object-oriented programming language can model a control system. Object-oriented languages are a good abstraction of the process control problem. Both the object-oriented language and the control system function by passing messages between objects. We plan to explore this approach in the future.

What is the desired average whisker length ?
(in inches 0. to 3.5)

3.0

We recommend reactor type B, which will you use ?
(A or B)

B

What is the desired average whisker diameter ?
(in microns, 0 to 12)

10

We recommend the manganese based catalyst, which
one will you choose ? (manganese or iron)

manganese

We recommend sieve size 20-25, what will you use ?
(25-32, 20-25, or 15-20)

20-25

How many times have you used the reactor growth plate ?
(0 or greater)

0

We recommend time-temperature profile A, which will you use ?
(A, B, C, or D)

A

We recommend that you use the following initial gas composition:
H2 = 80.0 %
CO = 5.5 %
N2 = 14 %
CH4 = 0.5 %

Will you use this ? (yes or no)

yes

We recommend that you vary the CO concentration according to
time-concentration profile A. What profile will you use ?

 (A, B, C, or D)

 A

Fig. 13. Dialogue with the expert consultant.

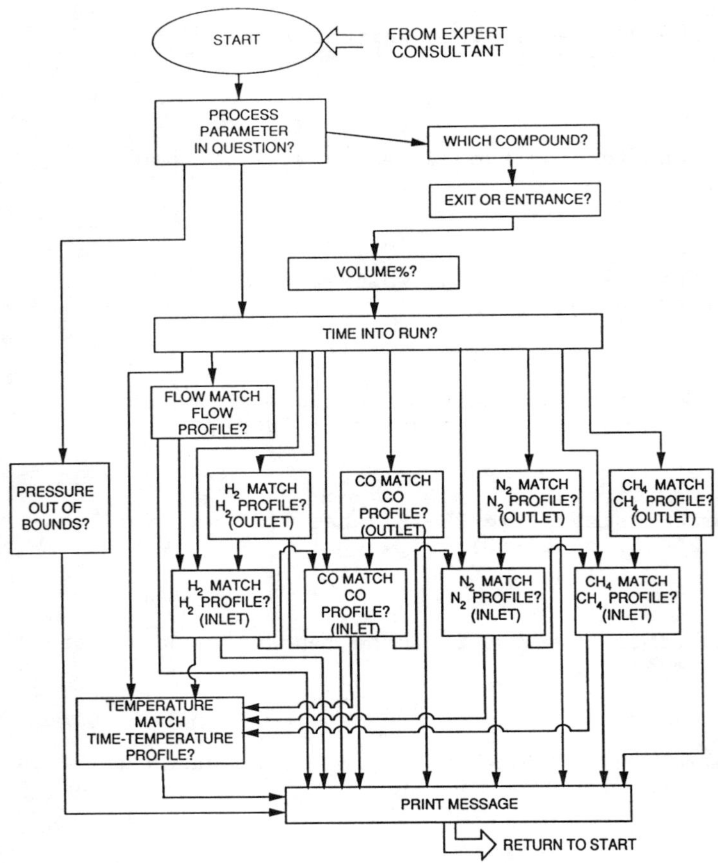

Fig. 14. The search space diagram for the expert control system.

Data Collection

Most of the rules used in our expert systems were obtained directly from the experts who design the whisker growth experiments and run the process. The rest of the rules were obtained from experimental data. Data from our gas chromatograph are read directly into the computer. These and other experimental data are used in a computer program to generate additional information that is stored in a database. A relational database management system is used to query the database and to produce tables that contain many different combinations of the stored information. The data in these tables are then plotted and analyzed in an effort to develop new rules. In some cases, we use least-square methods to fit the data and to produce new rules. In other cases, we use pattern-recognition techniques to find decision boundaries between data clusters. These boundaries are then used to produce new rules.

What process parameter do you wish to question ?
  (Temperature/ Pressure/ Total-Flow/ Gas-Composition)

Gas-Composition

Which component ? (H2/ CO/ N2/ CH4)

CO

Exit or Entrance Composition ? (Exit/ Entrance)

Exit

What is the volume percent ?

5

How many minutes since the run began ?

420

The volume percent is too high.

What is the inlet CO volume percent ?

4.6

What is the reactor temperature in degrees centigrade ?

1400

The temperature is OK.

Reduce the inlet composition to 2.9 volume percent and check
with me again in forty minutes.

Fig. 15. Dialogue with the expert control system.

Figure 17 is a cartoon showing this process for developing new rules.
Figure 18 shows the least-squares method used for developing a rule about the
whisker yield as a function of the number of uses of the growth plate.
Figure 19 shows the use of the perceptron algorithm (5) for finding a decision
boundary. In this case, we obtained data from our database for gas carbon
content (methane) for two given whisker diameter ranges, for a given gas
silicon monoxide content, for a given catalyst type, and for a given catalyst
particle size. A relational-database management system is an excellent tool
for obtaining stored data from a database in this fashion.

One problem with rule-based expert systems is that they tend to be brittle, that
is, they fail totally if they do not have a rule for every given situation. We are
analyzing the data carefully in an effort to develop as many rules as possible to
fill in the voids that occur in the system because of experts' incomplete knowledge. This is an effort to make our system less brittle.

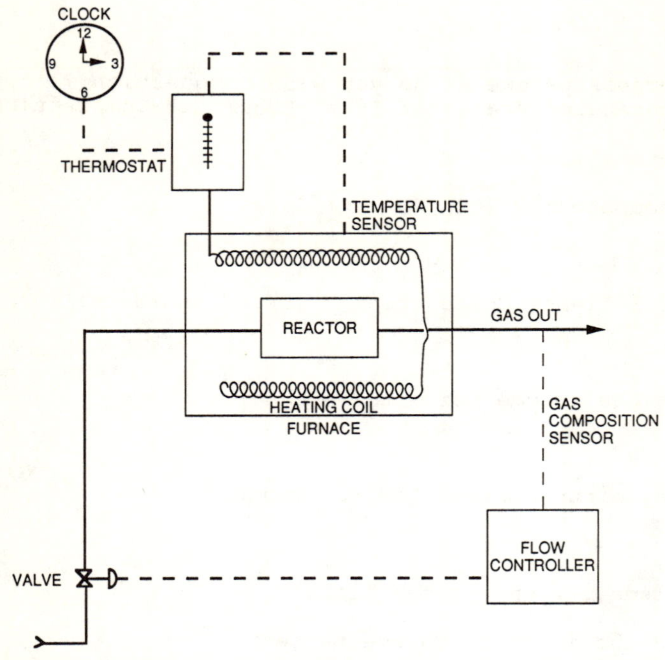

Fig. 16. A diagram of the model used by the object-oriented control program.

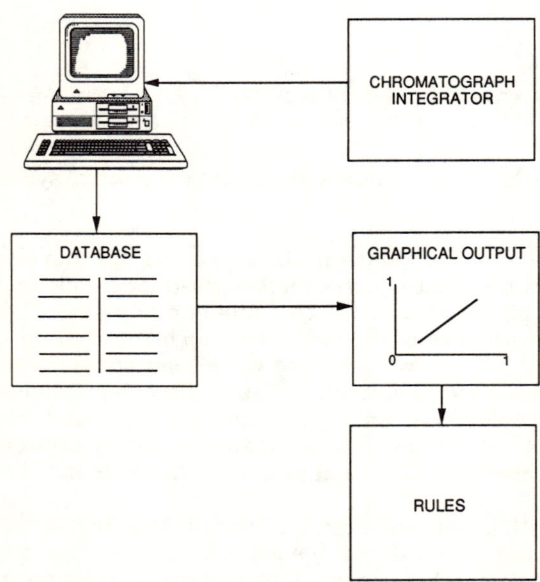

Fig. 17. A cartoon depicting the steps for developing rules for the expert systems.

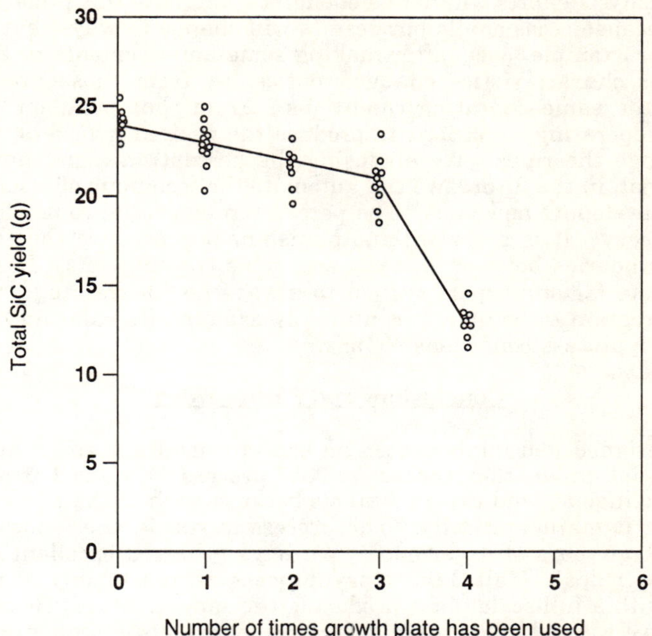

Fig. 18. Using the least-squares method for developing a rule about the number of times a growth plate should be used.

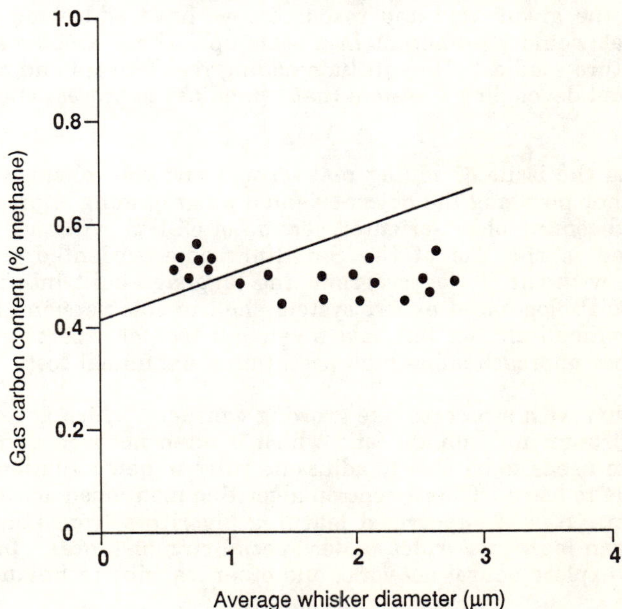

Fig. 19. Using a perceptron method for finding a decision boundary for developing a rule relating gas carbon content to mean whisker diameter.

Our data analysis efforts will not be enough to eliminate this problem entirely, however, because occasionally our results will change in ways that we cannot predict. For example, once, after making some improvements in the reactor, the whisker characteristics changed noticeably from those produced previously at the same operating conditions. From that point on, we had to change the operating conditions to produce the desired whiskers, that is, we had to change the rules. We are using the perceptron algorithm shown in Fig. 19 so that, in the future, we can automate the technique of testing existing rules and developing new ones. The perceptron algorithm comes from neural network theory. It is a reward-and-punishment procedure that establishes decision boundaries between linearly separable training sets. In the future, we would like for our expert system to always be "in training," that is, we want the program to be able to continually examine its rules and to develop new rules as process conditions change.

Conclusions and Future Work

We have designed and implemented an expert consultant and control system for a material production process. This process is a good candidate for artificial intelligence and expert systems because we have had great difficulty in modeling it mathematically. The process is run in the laboratory using rules based on years of experience. We have achieved excellent production using these rules. This laboratory process is reasonably simple when compared with a full-scale plant producing the same product. Because of this simplicity, we were able to design a working expert system, unencumbered by some of the complex problems of a full-scale process.

This program should be considered as a starting point only. To the best of our ability with the given time and resources, we have addressed some of the problems that would be inherent in a scale-up. These problems will be the subject of future studies. They include adding real sensors and controllers to our system and developing a system that can adjust to process changes or that can "learn."

In addressing the issue of adding real sensors and controllers, we are interested in further pursuing the object-oriented programming approach, possibly with the forthcoming object-oriented version of CLIPS. Another approach to be considered is the use of the Smalltalk object-oriented programming language (6) with the Prolog programming language built into it (7). If we can write the Prolog-based expert system shell in this version of Prolog and embed it into Smalltalk, we will have a valuable tool for expert system process control. Either approach offers high potential at a minimal cost.

We believe that with a process like growing whiskers, which is relatively sensitive to environmental changes and which is often not well understood, the expert system needs to be able to adjust its rules to new situations, that is, it should be able to learn. The perceptron algorithm mentioned above is a step in this direction. It is a supervised learning algorithm from neural network theory that can learn new rules under special circumstances. In the future, we intend to explore neural networks and other learning techniques further.

References

1.   W. J. Parkinson and D. E. Christiansen, "An Experimental Study of the Effect of Reactant Gas Mixing on Silicon Carbide Whisker Growth," Los Alamos National Laboratory report LA-10658-MS (February 1986).

2.   J. C. Giarratano, CLIPS User's Guide, Version 4.12 of CLIPS, Artificial Intelligence Section (Lyndon B. Johnson Space Center, October 13, 1987).

3.   G. F. Luger and W. A. Stubblefield, Artificial Intelligence and the Design of Expert Systems (The Benjamin Cummings Publishing Company, Inc., Redwood City, California, 1989).

4.   P. D. Shalek, D. S. Phillips, D. E. Christiansen, J. D. Katz, W. J. Parkinson, and J. J. Petrovic, "Synthesis and Characterization of VLS-Derived Silicon Carbide Whiskers," Proceedings of the International Conference on Whisker- and Fiber-toughened Ceramics (Oak Ridge, Tennessee, June 7-9, 1988).

5.   J. T. Tou and R. C. Gonzalez, Pattern Recognition Principles (Addison-Wesley Publishing Company, Reading, Massachusetts, 1974).

6.   Digitalk, Inc., Smalltalk/V - Tutorial and Programming Handbook (Digitalk, Inc., Los Angeles, California, 1986).

7.   G. L. Lazarev, "Prolog/V: Prolog in the Smalltalk Environment" (Dr. Dobb's Journal, November 1988, pp. 68-102).

A Meta-Controller

for Supervisory Process Control

Paul Nielsen, Vivek V. Badami, and James B. Comly

General Electric Corporate Research and Development
P.O. Box 8, Schenectady, NY 12345

Abstract

We describe a supervisory controller that incorporates both a procedural and rule-based language capable of responding to asynchronous real time sensor interrupts. This controller uses an English-like syntax that allows expressing process actions in both a sequential and an exception handling mode. The notion of time is explicitly incorporated into the syntax of the language. This system, referred to as a *Meta-Controller*, has been used to implement a supervisory control system for the growth of Gallium Arsenide crystals using the Liquid Encapsulated Czochralski process.

The Meta-Controller provides for consistent, repeatable growth of boules of Gallium Arsenide. It automates such process operator functions as monitoring and data logging. Using logical actuators, it changes control loop set points and control modes, and downloads new gain schedules to a digital controller based on changing conditions or timed events. It handles faults during the process through recovery or system shutdown.

This paper focuses on both the architecture for supervisory control of materials processing and an implementation on a physical puller using artificial intelligence techniques. It also describes the special demands of real time performance.

Expert System Applications in
Materials Processing and Manufacturing
Edited by M.Y. Demeri
The Minerals, Metals & Materials Society, 1989

Introduction

Advances in control theory have resulted in a great deal of autonomy in process control but have resisted complete automation. This is because the processes have operator level heuristics and experience built into them. The result of increased automation is operator inattention due to increased boredom [11], punctuated by brief moments of sheer panic. A goal of the Intelligent Processing of Materials Project is to alleviate the need for operator intervention by advances in several areas including sensors, process understanding, models, and controllers.

This paper discusses an artificial intelligence tool for supervisory control, the Meta-Controller (MC), which interacts with a digital feedback controller, and has been used to grow boules of GaAs. The MC implements the schedule followed by the operator and supplements it with world knowledge to anticipate deviations and take corrective actions. Most small scale deviations in the growth process are handled by the digital controller, but discontinuous or radical changes in the process (such as changing the control algorithm or accommodating device failures) are beyond its abilities.

In order to duplicate the high level reasoning used by operators, an intelligent system must be capable of performing the normal system operations by scheduling tasks to be performed as shown in the sample script in figure 2; logging events; maintaining a history of occurrences; detecting process deviations through unexpected or missing sensor reports; using temporal and causal reasoning to associate error symptoms with their actual cause; and finally controlling multiple processes simultaneously and in real time. By "real time" we mean that the system must be able to: follow the temporal evolution of the process underway, with suitable delays and waits; compute fast enough to match the time scale of the process to be controlled; guarantee a suitable answer within a specific time limit; and process asynchronous interrupts generated within the system or by external signals in a suitable manner.

The set of requirements is not complete, but is sufficiently broad that none of the presently available commercial expert systems are capable of dealing with them. Among the shortcomings of conventional AI tools are that they: are slow, have no clock, cannot handle asynchronous events, cannot communicate with other systems, and cannot deliver an answer in a limited time period. Recently there have been some AI systems which claim to have real time behavior (for example, ONSPEC Superintendent^{TM1}, G2^{TM2}, and R-TIME^{TM3}) but which were unavailable as this work began or inadequate for our purposes.

Background

The majority of GaAs crystals for computer chips today are grown using the Liquid Encapsulated Czochralski process (LEC). Essentially the process starts with gallium, arsenic, and boric oxide surrounded by argon in a sealed, rotating crucible. The mixture is heated to react the Ga and As to form liquid GaAs then maintained at just above the melting point, and a small seed crystal of GaAs is dipped into this melt. This crystal is slowly (over a period of about twenty to thirty hours) pulled out and the liquid solidifies into boule with the same crystal structure as the seed. (A extended descriptions of this process may be found in [9] or [10].) Faults which occur during this process include pinch off (bottom of crystal melting off), twinning (formation of a second crystal), thermal stress (causing poly-crystalline growth), and flash-out (melt solidifies too rapidly).

---

[1]Heuristics Inc.
[2]Gensym Corporation
[3]Talarian Corporation

1. Crucible preparation	6. Meltback
2. Charge preparation	7. Seed-on
3. Grower set-up and prechecks	8. Growth
4. Growth start-up	9. Termination
5. Synthesis	

Figure 1: Main Process Steps in LEC Process

Much work goes into developing process schedules, but the entire process is not well understood, and much of the time sensors only indirectly tell what is happening. The results depend on the skill of the operator. Most of an operator's time is spent adjusting parameters (set points), monitoring process parameters, documenting the current run, and making certain the state of the system matches the expected state. However, whenever a deviation from the expected behavior is detected, the operator must quickly assess what should have happened; what did happen; why it happened; and, most importantly, what to do about it. In addition, a quick (in the order of minutes) response to sudden deviations is usually essential, despite the fact that operators often have to divide their attention between several processes. Our goal is to capture those operator skills to produce consistent and repeatable boules.

Materials processing requires several different reasoning modes depending on the state of the process. During a normal growth run different schedules are used by the digital controller. There must be an awareness of the current state of events and processes as well as expectations of what might happen next. The applicability of the knowledge is tied to the current state of the process. For example, during the normal course of the pull losing contact between the boule and the melted GaAs is a highly erroneous situation (pinch off).[4] However, at the beginning of the run the boule may be started and burned completely off many times (seed-on) to obtain a more accurate temperature adjustment.

Portions of schedules for the LEC process and the seed-on stage of the process are given in figures 1 and 2 [7]. The schedule is written in a hierarchical manner. Within the seed-on step one sets a camera, repeatedly melts off the seed crystal to obtain a more accurate temperature reading, and then begins to pull the crystal from the melt. At each stage one knows what processes are active and anticipates various events which can happen. If only the desired events were to happen, the schedule could be followed blindly; however, this is never the case.

Our goal is not to duplicate work of developing the process schedule; but rather, when a deviation occurs the procedure should attempt to get back to the original schedule.

An Architecture for Supervisory Control

The MC has three major computation sequencing methods: procedural, temporal, and declarative; they are described in this section. Its architecture is shown in figure 3. Each of these may be used independently or in any combination with the others.

A critical difference between the MC and conventional rule based systems is that the MC is essentially a procedural language. Procedural languages provide focus, inheritance,

---

[4]In fact a slight change in the diameter of the boule is critical, without it burning off completely.

Precondition: The GaAs is completely molten and close to the dip temperature.
The crucible is rotating at W rpm.
The seed is stabilized in the oxide.

1. Make final camera adjustments

2. Lower the seed into the GaAs and take a reading

3. Wait for burn-off and record time (if after X minutes the seed does not burn off increase the heater temperature by Y)

4. Adjust the heater temperature based on burn-off time

5. Stabilize

6. Take a reading and redip

7. Stabilize

8. If seed burns off go to step 4

9. Start pulling at pull speed and make a log entry

10. When the length reaches Z take a reading and switch to run mode

11. Turn OFF the contact indicator

Postcondition: The seed is dipped into the GaAs and a stable crystal growth has begun

Exception handling:

1. Loss of heater power

2. Loss of coolant

3. Freeze occurs

4. Crucible vibration

Figure 2: Schedule Fragment of Seed-on Process

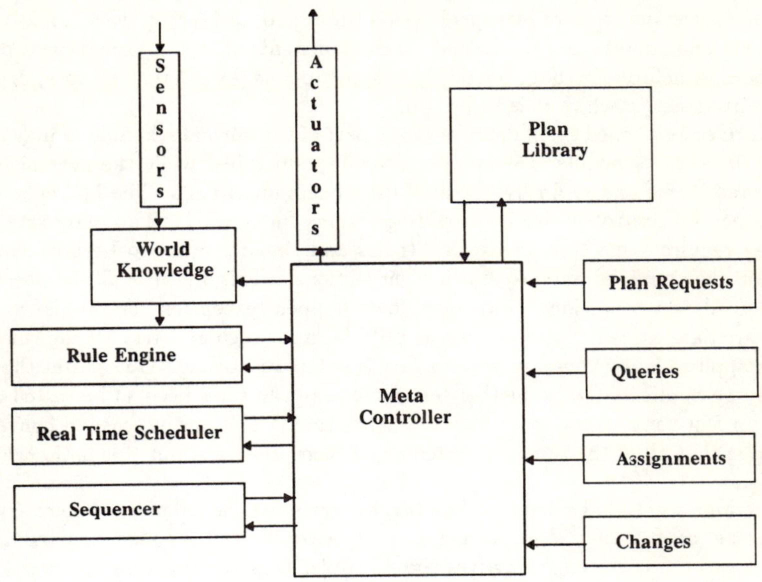

Figure 3: MC architecture

and context through scoping. Declarative languages require additional mechanisms or additional rule conditions to achieve these same effects. In the MC declarative statements and temporal statements are scoped in the same manner as variables in a procedural language. This means that only a subset of the rules are active at any time, but global rules (such as containment failure) are continually active and need not be constantly restated, as might be required with a conventional finite state machine approach. Rules may be posted at any time and will be retracted when leaving the scope of a plan or, in some cases, after they have fired once.

Procedurally is the way most computer languages operate, executing one statement after the other. The MC provides procedural semantics through Lisp. Statements in the control language compile directly to Common Lisp, and it accepts Common Lisp forms interspersed throughout its own syntax. Since these Lisp forms may be further compiled the MC is faster than an interpreted language without loss of flexibility. The standardization of Common Lisp allows the MC to port easily to a variety of machines. The hierarchical nature of the process schedule is given in a procedural fashion.

Declarative semantics provide the ability to react to changing conditions which may not occur at predictable intervals. For example, exception conditions such as crucible vibration may be detected and corrected, or a process may be followed until a signal arrives that it has finished. This ability is provided through a production system. Production rules have the form "if *conditions* then *actions*". The system works in a forward chaining manner. The conditions are statements about possible observations and may include variables. The actions are changes to the process, the numerical controller, or the beliefs of the MC which are made when the conditions hold.

Temporal reasoning is important to supervisory control. It provides information on the state of a continuous process which may not be detectable solely through sensor readings. In addition to keeping track of historical events through data logging, we can reason about future events and conditions with respect to clock instants. Clock information is provided in two modes, synchronous (busy waiting) or asynchronous (external interrupts) depending on the environment in which it is to be run.

Posted rules and timed events may occur as part of the normal schedule or may indicate deviations from the schedule. Deviations occur unpredictably, while the normal schedule is highly predictable and typically executed from beginning to end. The MC enhances the normal sequential control of Lisp by providing actions "in parallel". This allows statements which may require some time to execute (rules and timed events) to be posted without waiting for the previous ones to finish their actions. This in turn allows the user to anticipate or detect conditions which may never happen (exception conditions).

Rules are data driven. Systems such as OPS [5] must regularly test all the rules to see which are applicable; by contrast, when a fact is entered into the MC it causes those rules which may be applicable to be further tested. Not all the rules need to be tested on each iteration; in fact some of the rules may never be tested. The normal case in conventional rule-based is that all of the rules are tested at frequent intervals, but this is the worst case for the MC.

Rule actions and timed events are handled by wrapping the balance of the computation in a *continuation* function[5] and storing it on a scheduling queue, indexed by time and priority. When the correct time arrives the scheduler calls the function and waits for the next event. Rules and timed events may be posted at any time and will be retracted when leaving the scope of the subplan.

Finally, rules are integrated with temporal reasoning by providing an optional duration on the rule. This duration provides what Walther [12] called *potential time*. Semantically it means that if the rule is not activated within the given time, it will be retracted, and an optional, alternative course of action may be pursued. This allows time to be used as an additional logical constraint on the rule (i.e., if *conditions* **and** *time*, if *conditions* **or** *time*). It also allows the temporal relations of Allen's calculus [1] to be expressed with respect to observations (i.e, if *conditions* **before** *time*, if *time* **during** *conditions*).

Operational Features

This section discusses the factors other than program sequencing which need to be considered in a real time language. Not all actions are equally important, so when time is limited the most critical actions need to execute first. Beliefs change with time, and a result based on bad data is not acceptable. Reclaiming unused memory, while a minor annoyance in "normal" programming, causes the machine to stop processing at what might be a critical moment in control programming. Finally, even though a fast program is not necessarily a real time program, speed and predictability of response time are necessary considerations.

Priority

Actions do not have equal importance. For example, a request to shut down the apparatus should override a request to increase the setting of the heater because temperature is

---

[5]Continuation functions [4] allow a single process with common memory to be arbitrarily interrupted. In Common Lisp continuation functions maintain the same lexical scope as the environment in which they were created. Incomplete plans, waiting for some condition, would normally keep control of the processor or have their memory changed. Lexical scoping allows the processor to be freed without danger of losing variables, rules, and timed events.

a little low. The MC provides a priority mechanism for rules and timed events which aids in selecting the actions to execute first. These priorities are numbers which delay calculation until the request is posted. In this way the priority may be dynamically determined depending on the context.

At certain points in the program it is necessary to focus attention on critical problems and ignore some of the other possibilities. In addition to ordering execution by priority, the MC allows actions which fall below a given priority to be ignored. This is accomplished through a suspension. When the task finishes, lesser priority actions (if they are still applicable) may be resumed. Timed actions are recalculated to compensate for the time lost to the suspension. This feature is used by the operators to take manual control away from the MC while still allowing it to watch for exception conditions.

Sensor Validity

The certainty that a sensor reading is still correct decreases with time, and after a given amount of time has passed it is necessary to presume ignorance about the value, or take corrective action when the reading is critical. The time interval of validity of each value depends on the sensor, so it cannot be constant throughout the program. The MC has a primitive belief maintenance system which, among other things, keeps track of the validity interval associated with each sensor. After that interval is expired the reading will no longer be used.

Certain sensors are only valid for a limited number of readings. For example, a response from the operator might be required because of a missing sensor. If the operator fails to give the expected response, some corrective actions may be taken and the operator queried again. The operator's first response should not be reused as the new response, even though exactly the same query may be reposted. There are two solutions to this. The first is to clear that fact from memory before reposting the test; however, this requires additional code which may be overlooked causing programming errors. The second solution is to designate certain sensors as valid for a given number of uses. If the number of uses of that sensor exceeds the limit, the sensor is no longer believed. This solution requires that a sensor always has the same number of uses. Because this is not necessarily the case the MC supports both solutions.

Automatic sensor management is necessary for real time applications [9] [13]. It is not necessary to handle all of the data possible; in fact, processing slows considerably when one attempts to do so. Some facts may not be applicable in the given context, some of the facts may not have changed since the last reading, and some data may be ignored because of a suspension. The MC does not receive the actual values of the sensors; but rather, these are first preprocessed by another program and then converted to symbolic forms indicating ranges of values. That program, MICA[6] [3], has the ability to: selectively aggregate data; send data only when it falls outside a given range; send when it changes beyond a given threshold; send only by request; or not send at all until it receives a request to resume. The MC may send a request to the data filtering programs to begin receiving data from a sensor, get a new reading, or ignore some of the readings. Readings which confirm what the MC already believes will update the validity interval of that fact, but not repost the fact. In the future we hope to provide automatic mechanisms to look ahead and only post requests for data which may be used in the "near" future [8].

---

[6]Management Information Control Architecture

Memory Management

One of the problems with doing time critical applications on conventional hardware is garbage collection. The machine freezes while it reorganizes its memory to free space. Explicit garbage collection cannot be done because one cannot predict when an exception condition might occur. One can minimize the need for garbage generation caused by program execution through clever programming and avoiding the creation of dynamic structures at run time, but one cannot control the number or frequency of sensor readings. We have experienced cases where the machine would repeatedly go through garbage collection even though the program had not progressed.

The solution used in [9] is that sensor readings never become false, they simply belong to the past. This prevents storage reclamation, and works if the number of sensor values is expected to be small. However, in our experience new sensors are continually being introduced and, in addition to processing symbolic values for these sensors, numerical values are often used. (Storing numeric values without allowing memory to be reclaimed would lead to a machine crash.) The solution used by the MC is to log all readings in a file so that we can still reason about the past, but only store the last value of each logical sensor in memory. If a fact is no longer believed , but no new value has arrived it is explicitly marked as such without discarding the structure. By recycling memory structures only a number of facts equal to the number of sensors will ever be created.[7]

Speed Considerations

Because the general problem of unification (testing the applicability of rules) is NP complete, it is impossible, in the general case, to predict the time required to test a given number of rules. However, if one side of the unification is required to be constant, the speed of unification will only grow linearly with the size of this constant term. The MC enforces this by only allowing constants in the data. It still cannot guarantee a limited time for a given set of rules, without knowing the actual data, but in the future, since we know all possible sensor readings and data which might be produced, we hope to generate specific rules which require no variable instantiation. (This technique was described in [6].)

Knowing that the data will be constant offers a second speed benefit. One can tell by inspection of the rule triggers which tests will have to be made for unification to succeed, without looking at the data. This allows the rules to be open-coded, meaning these tests can be precomputed and compiled. Doing more of the work at compile time results in significant savings at run time.

Operator Interface

The final word in process control still rests with the human operator. Operators need the ability to take control from the MC easily, determine the state of the process, take actions, and return control to the MC. To do this they need to understand what the MC is doing and why it is doing it.

MC Plans are created by a process engineer or programmer using a structured English syntax. The syntax of the MC language is as close to the schedule syntax as possible. It provides temporal and causal statements to detect deviations from the schedule and provide corrective actions. If additional capabilities are required, one is free to use Common Lisp forms directly; however, this may degrade the real time performance and will confuse the

---

[7]This is not quite true since things other than logical sensors can be added to memory, but these are minimal compared to the sensor readings.

operators. The operators are not expected to write these plans, though they may as shown below, but they should be able to easily read and follow them.

A real time graphic display is maintained which shows the current state of the process as well as values of the logical sensors. It also shows the current overall plan and a history of the statements being executed.

When determining why the MC is taking some course of action it is necessary to see what is happening, what should be happening (when this is an exception condition), and what is needed to restore the desired state. This information is kept in a log file which contains a history of all readings and actions since the process began as well as the current plan. Tools are being developed to provide a better understanding of this file.

Operators can interact with this interface and start, stop, and suspend sequences through special features provided by the MC. In addition, because of the lack of distinction between interpreted and compiled code in Lisp, the operators may skip, alter, or introduce new plans while it is executing. To a lesser extent they may also alter the plan currently being executed by changing test conditions, times, or skipping statements.

Results

In this work we have created and demonstrated a supervisory control system for growing GaAs boules. It runs in real time and replaces most of the functions of human operators in monitoring and supervising digital feedback controls on a crystal growing apparatus. It has been used in growing two GaAs boules at the date of this writing, and other runs are expected to show further automation using it. The results for the Meta-Controller system, and for the crystal growth are described next.

Meta-Controller System

The system was built on a Symbolics 3640 Lisp Machine, and was ported to a Masscomp 5600 machine running a real-time version of Unix for testing against a simulator. It was then ported to a VAXStation 3100 which runs the software to control a Cambridge Instruments Model 356 Liquid Encapsulation Czochralski (LEC) crystal puller.

Scripts for plans were written using the MC's language described above. Its grammar is as close to a language that is natural for writing sequential plan schedules as possible. Is is supplemented with a temporal and causal syntax, and with declarative statements, which allow detection and handling of process exceptions.

A graphical user interface allows the operator to see the execution status of the system, values of logical sensors and actuators, and to interact with the system using dialogue boxes and menus.

The Masscomp and VAX implementations run on Lucid Common Lisp; porting the Meta-Controller itself from the Symbolics was straightforward and required rewriting about 5 functions. More major rewriting was required for the interprocess communication between the Meta-Controller and the rest of the system; and the graphic operator interfaces had to be rebuilt.

The details of the overall Real Time Control System which runs the LEC apparatus is described in more detail in [2].

Crystal Growth

The system has been used to automate sections of the full crystal growth protocol for GaAs using the LEC apparatus. Two 4 inch diameter boules have been grown from 8 kg charges using the system where plans have been implemented. Portions of the Synthesis

187

and Meltback steps have been implemented, and the Seed-On step which requires the most manual operation is almost completely automated. Work is under way to complete implementation of plans for the remaining phases of crystal growth.

Conclusions

This paper demonstrated an architecture for real time control. The work presented here should be easily implementable by a good Lisp programmer with AI experience and suggest features to look for when selecting an expert system for process control.

The MC is designed to be fast, portable, and flexible. Since it is only one aspect of a much larger project, process methods are constantly being refined and schedules are changed to accommodate them. At this writing there are insufficient sensors and actuators to completely automate the process, so communication between the MC and the operators must be easy and clear. As new sensors and actuators are developed they must also be accommodated.

In the future we hope to explore qualitative models of the underlying physical process. These can be used as process simulators to check the completeness and correctness of the process plans. Rather than replanning during process execution, qualitative models use all possible symbolic values for the sensors at all stages of the process to exhaustively determine all possible states and transitions of the process. This information can be compiled and used during the run to anticipate faults and provide more robust methods for system recovery or shutdown.

REFERENCES

[1] James F. Allen. "Maintaining knowledge about temporal intervals". In R.J. Brachman and H.J. Levesque, editors, *Readings in Knowledge Representation*. Morgan Kaufman Publishers, 1985.

[2] Vivek V. Badami, Paul Nielsen, and James B. Comly. "A real-time, rule-based supervisory controller for process automation". In *Proceedings of the 1989 TMS Fall Meeting*, Indianapolis, IN, 1989.

[3] Bruce Bernstein. MICA. Unfortunately a description of MICA is unavailable at the time of this writing. More information may be obtained from the author at GE, CRD.

[4] Eugene Charniak, Christopher K. Riesbeck, Drew V. McDermott, and James R. Meehan. *Artificial Intelligence Programming*. Lawrence Erlbaum Associates, Hillsdale, NJ, 1987.

[5] C. L. Forgy and J. McDermott. "OPS, A domain-independent production system language". In *Proceedings of the Fifth International Joint Conference on Artificial Intelligence*, pages 933–939, Cambridge, MA, 1977.

[6] F. D. Highland and C. T. Iwaskiw. "Knowledge base compilation". In *Proceedings of the Eleventh International Joint Conference on Artificial Intelligence*, pages 227–232, Detroit, MI, 1989.

[7] Timothy A. Mancour. *8KG Process Description*. Spectrum Technology, Inc., January 1988. The process descriptions used here are for illustration only and do not necessarily correspond to those given in this manual.

[8] T. Murayama, K. Kushima, and S. Ishigaki. "An inference mechanism suited for real-time control". In *Second International Conference on Industrial and Engineering Applications of AI and Expert Systems*, Tullahoma, TN, 1989.

[9] William J. Pardee and Barbara Hayes-Roth. "Intelligent real time control of material processing". Technical Report Research Report 1, Rockwell International Science Center, Palo Alto Laboratory, 1987.

[10] K. Riedling. "Autonomous liquid encapsulated Czochralski (LEC) growth of single crystal GaAs by 'intelligent' digital growth". *Journal of Crystal Growth*, 89:435–446, 1988.

[11] Peter Senker, Joe Townsend, and Joanna Buckingham. "Working with expert systems: Three case studies". *AI and Society*, 3(2):103–116, 1989.

[12] Eckart Walther, Vivek V. Badami, James B. Comly, Paul Nielsen, and Van-Duc Nguyen. "Real-time temporal and causal reasoning for intelligent control". In *Proceedings of the Industrial and Engineering Applications of Artificial Intelligence and Expert Systems*, Tullahoma, TN, 1989. Association for Computing Machinery and The University of Tennessee Space Institute.

[13] Richard Washington and Barbara Hayes-Roth. "Input data management in real-time AI systems". In *Proceedings of the Eleventh International Joint Conference on Artificial Intelligence*, pages 250–255, Detroit, MI, 1989.

MICAPP: AN EXPERT SYSTEM FOR PROCESS PLANNING

M. S. Gouda

Senior Engineer
Detroit Edison
1159 Falcon
Troy, Michigan 48098

K. S. Taraman

Dean of Engineering
Lawrence Technological University
21000 W. Ten Mile Road
Southfield, Michigan 48075

Abstract

Expert systems are needed for the creation of a comprehensive process plan.
These intelligent systems are computerized and often called Computer Aided
Process Planning (CAPP) systems. In an intelligent process planning system,
a knowledge base, a data base, and a logic inference form the basis of an
intelligent system. Rules and facts are represented by several intelligent
languages such as LISP and PROLOG.

A comprehensive survey was conducted to identify these CAPP systems utilizing
the expert system approach. Other systems using different approaches were
also included, i.e., the Generative, Semi-generative, and variant approaches.
Their limitations and/or shortcomings are discussed. This paper identifies
128 CAPP systems. These CAPP systems are listed by the type of input method
which is used to describe the design specifications to the CAPP system. This
paper illustrates the results of the comprehensive survey, including the
computer hardware, part-type, input of data, processing logic, selection
criteria (or models), the output, and the developer. Fourteen countries have
been identified.

Expert System Applications in
Materials Processing and Manufacturing
Edited by M.Y. Demeri
The Minerals, Metals & Materials Society, 1989

Introduction

The purpose of the process planning function is to identify steps by which a part can be efficiently produced at a minimum cost. The manual process planning function is known to be tedious and time consuming. It is estimated that about 85% of a product's cost is established during the design phase (10,17). However, this cost is significantly reduced if the manufacturing personnel are involved during that phase.

It was reported that the computers will be used to generate over 50% of the process plans for simple or assembled parts (7). Computer Aided Process Planning (CAPP) methods can overcome the shortcomings of the manual process planning. There are three main classes of Computer Aided Process Planning methods. These are the variant, generative and intelligent systems. The essential elements of any automated process planning are illustrated in Figure 1.

The objective of the variant type is to create a process plan for a few parts by retrieving an existing plan and/or modifying it as necessary. The generative approach is known to provide an optimal fabrication sequence through a series of enhanced and/or sophisticated decision logic, rules or algorithms (5,6,8). More refined process decision logics are being developed and expert systems are gradually applied into CAPP environment. A knowledge base, a data base and a logic inference form the basis of an intelligent system. Table I contains a list of the surveyed systems including their type of planning approach, i.e., generative (G), semi-generative (SG), variant (V) and the newly developed field of artificial intelligence (AI). The survey identified 128 systems with different performances and limitations (13).

Representation of Workpiece and Input Mode

Design specification of a workpiece such as dimensions, tolerance and surface integrity are represented in many different ways. The input format of the design specification to a generative process planning can be performed through one of the following three generic methods:

1. The Code Method

In this method a unique code is developed and utilized by the user. This code is entirely different from the Group Technology (GT) code which is used in the variant systems. These types of codes are more detailed and comprehensive.

2. Descriptive Languages or Methods

A special language can be used to provide detailed part information at the input phase of process planning; language similar to solid modeling with primitive shapes. Some of the descriptive languages or methods are: cell constructed geometric modeling language, feature and work element language, English-like descriptive language, solid geometrical modeler language, boundary representation solid modeler language and elementary machine surfaces language.

3. Computer Aided Design (CAD Models)

Some computer aided process planning systems have the capabilities to interact and/or interface with CAD models and accept their output data as input information for processing. This approach is more desirable for

192

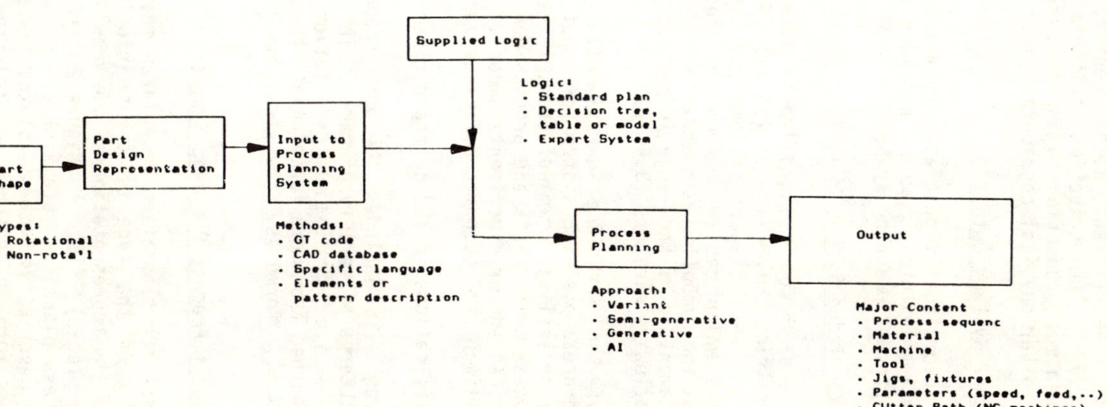

Fig. 1 Essential elements of automated process planning (13,14,15).

computer aided process planning because of its direct interface with a CAD
system and without human intervention (5).

Processing Logic and Selection Criteria

The objective of a decision logic in a CAPP system is to match the process
capabilities with design specifications in an optimal way. Generally, the
most common decision logic can be classified as one of three methods:
decision tables (21), decision trees (1) and, most recently, the field of
artificial intelligence (9,22). The selection criteria for process
sequencing may be based on minimum production time, cost, combination of
both, economic, management or other physical considerations.

CAPP Systems Output

Seven possible features are automatically generated by a process planning
system and are known as automation features. Table I provides the automation
features for most generative process planning systems.

Fundamentals of a Knowledge System in Process Planning

Expert process planning systems showed a significant decrease in their
reliance on the sole problem solving theory domains (16,18). Expert systems
are also known as knowledge systems. Expert systems possess and involve
several distinguished features which can be applied to process planning
(16,23). These systems include the symbolic representation, heuristic search
and symbolic inference. Fundamentals of knowledge in process planning will
include one or more of the following elements (3,4,12):

1. Representation of the workpiece and the transfer of this
    technology to a process planning system for input.
2. Machinability, producibility, or manufacturability of parts.
3. Manufacturing process constraints and process monitoring.
4. Analytical, empirical and/or experimental models.
5. Optimization techniques and economic models as applied to metal
    cutting.
6. Machine tool specifications including their capabilities.

Knowledge in process planning also includes the understanding of shop
practices, performance and patterns of variety of parts. This knowledge is
acquired through practical means, i.e., profession, association, acquaintance
or mentors. It can also be learned from more formal means through schooling
and continuing education, i.e., formulas, theories, axioms, laws and
principles (13).

Review of Expert Process Planning Systems

Expert process planning systems are classified into three main categories.
The classification is based upon the type of input information to the
knowledge system regardless of the representation of the knowledge (semantic
network, OAV, rule, frame logic) (16). Figure 2 illustrates the
classification of expert process planning systems which includes: (a) the
knowledge systems with code or descriptive method, (b) knowledge systems with
representation of parts in the form of features and (c) knowledge systems
with CAD interface.

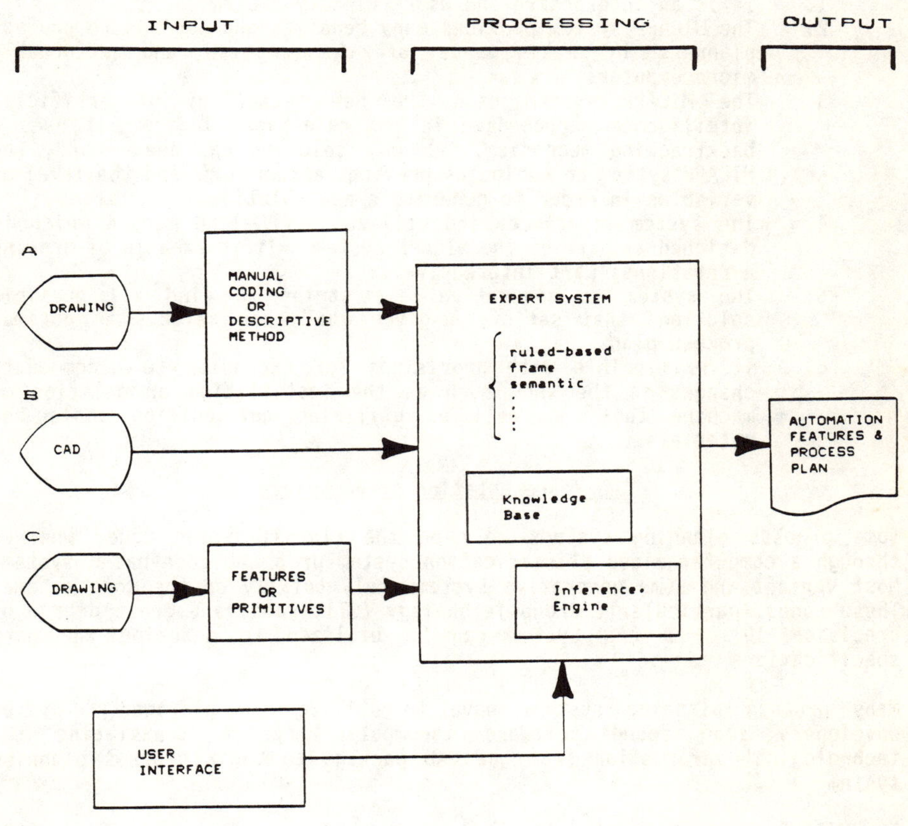

Fig. 2   Classification of expert process planning systems (13).

<u>MICAPP System</u>

An intelligent computer aided process planning system for rotational parts
requiring turning operation has been developed. This Micro-based Computer
Aided Process Planning System is named MICAPP. The structure of the MICAPP
system is shown in Figure 3. The significant design philosophy of the MICAPP
system is summarized as follows (13,14,15):

1.   It is an interactive and user-friendly system.
2.   The MICAPP system provides many benefits and savings to process
      planners by making use of the popular and affordable
      microcomputers.
3.   The MICAPP system uses the new technology of Artificial
      Intelligence represented in its rule-based logic. It uses a
      backtracking mechanism. Once a solution has been found, the
      MICAPP system re-evaluates previous assumptions and the level of
      variables in order to generate a new solution.
4.   The system interfaces and utilizes a CAD-like module which is
      designed as part of the MICAPP system. It is capable of drawing
      a rotational part interactively.
5.   The system is generative. It tries to find all possible
      solutions that satisfy a given goal then selects an optimal
      process plan.
6.   It is flexible with provisions for expansion to accommodate
      changes in the shop such as the installation or deletion of
      machine tools as well as utilizing new cutting tools and
      materials.

<u>CAD Representation of Workpiece</u>

Some process planning systems rely on the classification code, whether
through a computer-aided classification system or a CAD code-based system.
Most variant and some generative systems rely totally on the code scheme.
These codes, particularly Group Technology (GT), require a great degree of
precision (19). A CAD System can be utilized to describe the part
specifications.

Many process planning systems have to utilize CAD packages. System
developers soon found software incompatibility in translating the
technological information from the CAD package to their process planning
system.

The MICAPP system attempted to resolve these concerns by developing a CAD-
like system without relying on the use of external stand-alone CAD package.
The CAD portion of the MICAPP system can draw cylindrical, tapered, threaded
and grooved surfaces for both internal and external operations. An example
of a typical part generated by the MICAPP system is shown in Figure 4. A
typical input consultation of a workpiece is also shown in Figure 4.

<u>Knowledge Representations</u>

The knowledge required to produce a process plan is divided into eight types
in the MICAPP system. These types are listed as follows (11,20,24):

1.   Stock material knowledge.
2.   Cutting tool knowledge.
3.   Selection of Optimal Conditions.
4.   Selection of machining parameters knowledge.

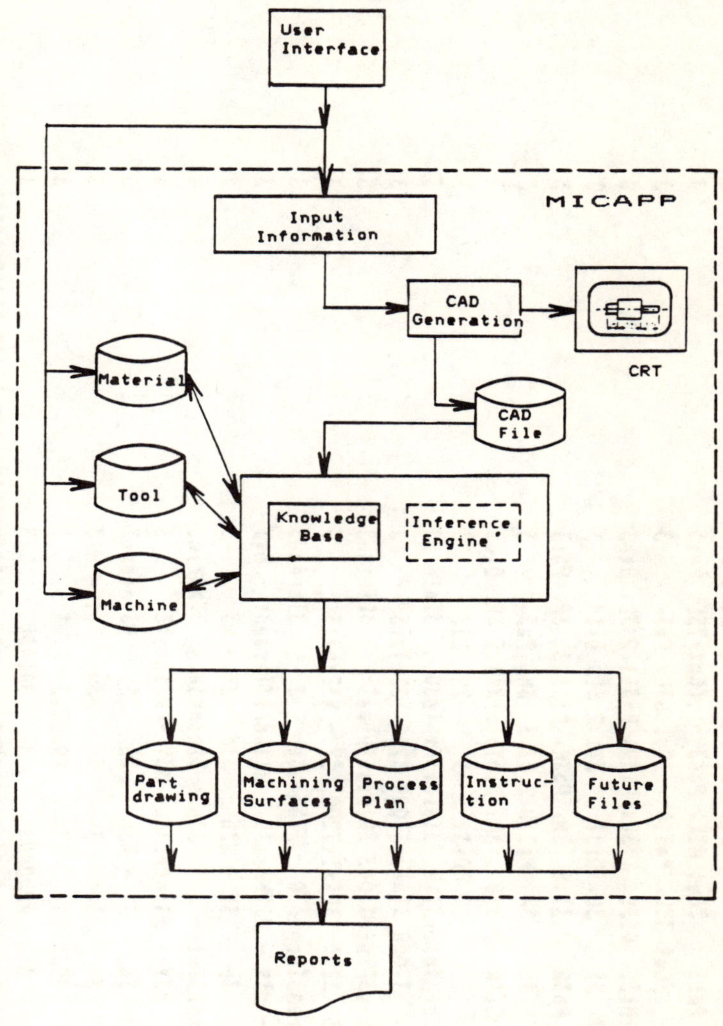

Fig. 3   The structure of Micro-knowledge-based Computer
Aided Process Planning (MICAPP) System (13).

```
ZDDDDDDDDDDDDDDDDDDDDDDDDDData Input -- Cut SpecificationsDDDDDDDDDDDDDDDDDDDDDDDDD?
3Enter specifications for each cut. RETURN after all cuts have been described. 3
3       ZCut 20?            ZCut 50?            ZCut 80?                          3
3ZCut 13L. Pos3     ZCut 43L. Pos3     ZCut 73L. Pos3                            3
33L. Po33     ZCut 33L. Po310    ZCut 63L. Po3        3                          3
331     3L. Di3L. Di38     3L. Di3R. Po311.2 3L. Dia3                            3
33L. Di32.7 31    3L. Di31     311.2 3L. Di3        3                            3
330     3R. Po3R. Po31.8 3R. Po3R. Di31.3 3R. Pos3                              3
33R. Po37     38     3R. Po310.2 31.3 3R. Po3        3                           3
333     3R. Di3R. Di310   3R. Di3Toler311.2 3R. Dia3                            3
33R. Di31    31    3R. Di31    3     3R. Di3        3                            3
332.7 3Toler3Toler31.8 3Toler3Finis30    3Tolera3                               3
33Toler3       3.008 3Toler3.009 3200 3Toler3        3                          3
33      3Finis3Finis3.003 3Finis3Threa3       3Finish3                          3
33Finis3125   3100  3Finis3200  3Y - N3Finis3        3                          3
33125  3Threa3Threa330    3Threa3y    3      3Thread3                           3
33Threa3Y - N3Y - N3Threa3Y - N3Thd/i3Threa3Y - N:3                             3
33Y - N3n     3n    3Y - N3n     314   3Y - N3        3                          3
33n    3Inter3Inter3n    3Inter3Inter3n     3Thd/in3                           3
33Inter3Y - N3Y - N3Inter3Y - N3Y - N3Inter3        3                           3
33Y - N3n     3n    3Y - N3nn    3n    3Y - N3Intern3                           3
33n    3     3     3n     3     3     3n     3Y - N:3                            3
33     3     @DDDDD3     3     @DDDDD3     3     3                               3
3@DDDDD3     3     @DDDDD3     3     @DDDDD3     3                               3
@DDDDDDD@DDDDDDDYDDDDDDDDDDD@DDDDDDDYDDDDDDDDDD@DDDDDDYDDDDDDDDDDDDDDDDDDDDDDDDDDDDDY
```

Fig. 4-a   Input for the MICAPP example.

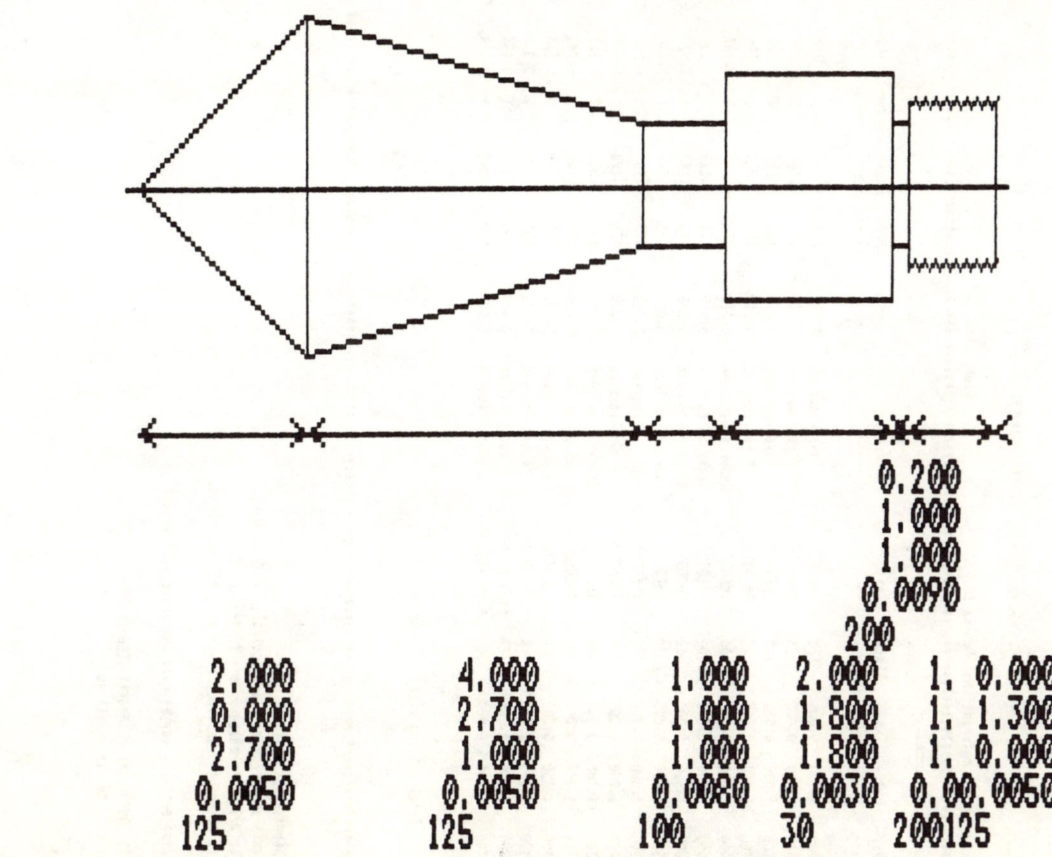

Fig. 4-b MICAPP-CAD part illustration.

Process Plan

Prepared by: Saied Gouda, PE.

Material: Hot Rolled 1040   Stock Size: 3.5
****

Op.No.	L.Dia.	R.Dia.	Tol.	Finish	Tool	Process	ipr	Speed	Depth	#pass	Time,min
1	0.000	2.700	0.005	125	TNEG322	Taper/Conical	0.010	173	0.187	5	1.234
1	0.000	2.700	0.005	125	TNEG322	Taper/Conical	0.010	173	0.184	1	0.408
2	2.700	1.000	0.005	125	TNEG322	Taper/Conical	0.010	173	0.187	4	2.572
2	2.700	1.000	0.005	125	TNEG322	Taper/Conical	0.010	173	0.083	1	1.119
3	1.000	1.000	0.008	100	TNEG322	Heavy rough turn	0.010	173	0.187	6	0.907
3	1.000	1.000	0.008	100	TNEG322	Heavy rough turn	0.010	173	0.128	1	0.151
4	1.800	1.800	0.003	30	TNEG332	Heavy rough turn	0.020	173	0.187	4	3.320
4	1.800	1.800	0.003	30	TNEG332	Heavy rough turn	0.007	173	0.102	1	0.830
5	1.000	1.000	0.009	200	TNEG332	Grooving	0.020	173	0.187	6	0.091
5	1.000	1.000	0.009	200	TNEG332	Grooving	0.020	173	0.128	1	0.015
6	1.300	1.300	0.005	200	TNEG666	Heavy rough turn	0.030	173	0.500	2	0.163
6	1.300	1.300	0.005	200	TNEG666	Heavy rough turn	0.024	173	0.100	1	0.082
6	1.300	1.300	0.000	0	AKUS	Rough Thread	0.071	250	0.020	7	0.133
6	1.300	1.300	0.000	0	AMIR	Finish Thread	0.071	250	0.001	1	0.019
7	1.300	0.000	0.005	125	Face tool	Rough facing	0.002	600	0.100	1	0.184
7	1.300	0.000	0.000	125	Face tool	Finish facing	0.002	800	0.060	1	0.184

        Total cutting time: 11.41

****

Unless otherwise specified :
1. Tolerance on two/three place decimal, .005 inch
2. Chamfer all sharp edges 0.06 x 45 degree.
3. surface roughness, arithmatic average, 125 microinch.

The following machines may be used for these operations:

Name: WHITE CONSOL Inc   Model: 60CH-cnc
Name: Monark   Model: Medallion

Fig. 4-c The process plan for the part shown in Fig. 4-b.

October 6, 1989

NO.	FULL NAME	Acronym	App-roach	Computer Hardware	Rotational	Non-rot	Cad	Code	Other	Interactive	Other
					PART TYPE		Part Description			Mode	
34.	Computer Aided Tool Selection	COATS	AI	VAX 11/730				X		X	
43.		CUTTECH	AI					X			
50.	EXpert Computer Aided Process Planning Sys.	EXCAP	AI	VAX 11/750	X				X	X	
54.	Generator for p'ng sequ. of m/g cuts of mech	GARI	AI	HB-68		X			X		
62.		INTELLCAPP	AI		X			X			X
64.	Know how and knowledge Ass. Prod. Plan'g Sys.	KAPPS	AI	Mini-Comp	X	X	X		X	X	
78.	An Intregrated Intelligent Design and PPS	MICROPLAN	AI	Mini-Comp	X		X			X	
95.	The PROPEL Process Planner	PROPEL	AI	UNIX3-160	X	X			X	X	
103.	Manufacturing System Design Theory	SAPT-EXPERT	AI	Mainframe	X	X		X	X	X	
106.	Semi-Intelligent Process Selector	SIPS	AI	Symb-lisp		X			X	X	X
116.	Technostructure of Machining	TOM	AI	VAX-11		X	X		X	X	
119.	Intelligent Reasoning for Process Planning	TURBOCAPP	AI	IBM PC	X		X	X		X	
126.	Unpublished Thesis	XCUT	AI			X					
127.	Expert Process PLANning Environment	XPLANE	AI	VAX-II		X			X	X	
5.		AMP	G	Mainframe							
6.	Automatically Processing of Mfg. Info.	API	G	Mainframe		X			X	X	
7.		APLAN	G		X	X					
8.	Automatic Process Planning & Selection	APPAS	G	PDP-10		X		X			X
9.		AUTAP	G		X	X	X			X	X
10.	AUTOmactic Computer Assisted Planning sys.	AUTOCAP	G	PDP-8	X				X	X	
12.		AUTOPLAN	G	Mainframe	X		X	X			
14.	AUTOmated PROcess Planning System	AUTOPROS	G	Mainframe							
15.		AUTOTECH	G	ES-1040	X		X				X
17.	Boeing Generative Process Planning	BGPP	G	Proprietary	X	X	X			X	
21.		BPT	G			X					
22.	Computer Aided Design & Computer Aided Mfg.	CADCAM	G	IBM 370/158		X	X			X	
23.	Computer Aided Machining Parameters Selection	CAMPS	G	Mini-comp	X			X		X	
25.		CAPES	G		X	X					
26.		CAPEX	G		X	X					
29.	Computer Aided Planning SYstem	CAPSY	G	Mainframe	X					X	
30.	Computer Aided Routing	CAR	G	Mainframe					X	X	
31.		CAWP	G	Mainframe							
32.	Computer Aided Integrated Mfg Sys/PROduction	CIMS/PRO	G	FACOM 230/35		X					
33.	Computer Managed Process Planning	CMPP	G	Univac/IBM	X		X				X
35.		COBAPP	G		X			X			X
38.		COMPAC	G	Mainframe							
41.	Computerized Production Process Planning	CPPP	G	Mainframe	X			X		X	
42.		CUTPLAN	G		X	X		X			
45.	Decision CLASSification information system	DCLASS	G	Mainframe	X	X		X		X	
46.		DISAP	G		X	X	X			X	
47.		DREKAL	G		X	X			X	X	
48.		ENCEPE, 111GNC	G	Proprietary	X	X					
49.	EXtension Automatic Programming Tools	EXAPT	G		X	X			X	X	X
53.	Feature Recognition & Expert Process Planning	FREXPP	G			X					
58.		GLEDA	G		X						
59.		GTIPROG	G		X	X					
66.	Layout Est., Tool'g & Setup-Multisp Bar auto	LETS-MB	G	Mini-Comp		X		X		X	
67.		LOCAM	G		X	X		X			X
68.	Machining oriented Automatic Decision sys.	MAD	G			X			X		X

| | OUTPUT | | | | | |
| PROCESS | | | Inter-Face with CAM | DEVELOPER | | | |
Logic Type	Selection Criteria	Automation Features		Name	Country	Known year	Reference No.*
ES	W	d		PISA	ITALY	1986	66
ES		a,d,f		METCUT	US	1987	161
ES		a,f		UMIST	GB	1984	35
ES	DR	a		ADEPA	FR	1980	39
ES				CIMTELL-	US	1985	158
ES	DR	a,c,d,e,f		KOB U	JAP	1986	84,85
ES	CF	a,d,f		UOI	US	1986	124
COP	CA	a,d		ITMI	FRAN	1986	157
ES	DR	a,c,d		BELG.UNI	YUG	1987	110
B&B	LC	a		UOM	US	1986	114
ES	DR	a,f,g	x	UOT	JAP	1981	104
MR	C	a,d,f		PSU	US	1986	166
ES		a		UOI	US	1986	147
ES	C	a,d		UO TWENT	NETH	1986	53,54
				IBM	US		165
DTr	T&C	a,g	x	UOT	JAP	1979	106
	T&C	a,b,c,f		GEBER	WG		59
DTr	T	a,d,f		PU	US	1977	174
DTa		a,b,c,d,f,g	x	WZL	WG	1978	57
DM		a,b,c		UMIST	GB	1979	49,59
MR	T&C	a,b,c,d,e		METCUT	US	1971	163,169
		a		NAKK	NOR	1972	
DTa	T	a,b,c,d,e,f			EG	1979	152
DTr		a,b,f		BOEING	US	1978	50
	T&C	a,b,c,f		VEBIKMS	EG		59
DTa	T	a,d,f		VPT	US	1980	24,26
DTa	T	b,d,f		PU	US	1982	133
	T&C	a,c,f		MW	GB		59
	T&C	a,b,c,d,f,g	x	EXAPT	WG		59
DTa	T&C	a,b,c,e,f,g	x	IWF-B	WG	1978	144
SP	T	a,c,f		KOBE-U	JAP	1973	77
					WG		165
		a,c,d,g	x		JAP	1980	83
DM		a,b,c,d,g	x	U.TECHNO	US	1982	70,129,164
		a		PU	US	1978	125
					EG		165
DM		a,c,d,f		UTRC	US	1977	44,45
ES	T&C	a,c,f		METCUT	US	1987	59,158
DTr		a,b,c,d		BYU	US	1974	4
DTa	T&C	a,b,c,d,f,g	x	WZL	WG	1976	59
	C	a,b,c,d,e,f,g	x	IFWTU	WG	1981	59
		a,b,f		SU	SW		59
		f,g	x		WG	1969	22,169
						1984	95
		a,c,f		IOT	HU		59
	T	a,f		IOT	HU		59
DTa		a,c,d		TIPNIS	US	1981	137
SP		a,b,c,f		LOGAN	GB	1982	101
DTa		a,c,d,g	x	MMMC	JAP	1974	152

| NO. | FULL NAME | PLANNING SYSTEM | | | PART TYPE | | INPUT | | | | |
| | | Acronym | App-roach | Computer Hardware | Rota-tion-al | Non-rot. | Part Description | | | Mode | |
							Cad	Code	Other	Inter-active	Other
70.	Makino Automatic Program'g and Process'g Sys.	MAPPS	G	MAINFRAME		X		X		X	X
71.		MAYNARD	G								
73.	An auto program'g sys. for lathe op., MELTS	MELTS	G		X			X			
75.		MICON	G		X	X					
77.	Microcomputer based Precess Planning System	MICRO-GEPPS	G	IBM PC	X			X		X	
81.		MITSUBISHI	G	Proprietary	X	X					
83.	Machining Kernel Software	MKS	G			X	X		X	X	
84.	Mechining Center Operation Planning System	MOPS	G			X	X		X	X	
88.	Overall Planning & Optimizing of M/g Process	OPOMP	G	Mini-Comp	X			X		X	
90.		PICAP	G	Mainframe	X		X				X
91.	Proc's & Op. Plan'g sys Based on Usr Lang/Res	POPULAR	G	Mainframe	X	X	X			X	
92.	Process Planning by Information Concept	PPINC	G	Mini-Comp					X	X	X
94.		PRIKAL	G			X					
96.		PROPLAN	G		X	X					
97.		PS-SYSTEM	G		X	X					
98.		RATIBERT	G		X	X					
99.	Adaptive Control Module	ROUND	G	PDP11/73	X					X	
105.		SIB	G		X	X		X			
107.		SISPA	G		X	X		X			
109.	Sequential and Tool Oriented Process Plan'g	STOPP	G	Mini-Comp		X			X	X	
114.	Totally Integrated Process Planning Sys.	TIPPS	G	Mini-Comp		X			X	X	
118.		TRAUPROG	G	Mainframe	X		X				
120.	Information center for cutting data (INFOS)	TURN	G		X	X					
122.		VERDI	G		X	X					
124.	Westinghouse Integrated Computer Aided PP	WICAPP	G	Mainframe	X	X		X		X	X
128.	eXperimental Planning System	XPS-1	G	Mainframe	X	X			X	X	
2.	Automated Coding and Process Selection	ACAPS	SG	PDP-11	X	X		X		X	X
16.	Beijing Institute of Mechanical Eng.	BGCAP	SG	MICRO	X	X			X		
18.	Beijing Institute of Aeronautic & Astron	BHCAP	SG	MICRO	X				X		
19.	Beijing Institute of Technology CAPP	BITCAPP	SG	MINI/MICRO	X				X		
28.	Computer Aided Planning System	CAPS	SG	CDC 7600	X			X			X
37.	COmputer GEnerated Route Sheets	COGERS	SG	Mainframe	X	X		X		X	X
40.	Company-Oriented Ruled-Based Expert CAPP	CORECAPP	SG	Micro	X			X		X	
44.		DATASAAB	SG		X	X				X	
55.	GENerative PLANing system	GENPLAN	SG	IBM 370/168	X	X		X		X	
56.		GENTECH	SG		X						
61.	Interactive Computer Aided Process Planning	ICAPP	SG	Mini-Comp		X			X	X	X
101.	Rotating Parts Operation	RPO	SG	Mini-Comp	X		X	X		X	X
104.		SGPP	SG	Mainframe							

| | PROCESS | OUTPUT | | DEVELOPER | | | |
Logic Type	Selection Criteria	Automation Features	Inter-Face with CAM	Name	Country	Known year	Reference No.*
MR	T	a,c,d,f,g	x	MAKINO	JAP	1985	135
						1980	177
		a,b,e,f,g	x	TAKEYAMA	JAP	1973	50
	T	c,f		TUB	HU		59
DTr		a,d,f		PSU	US	1985	165
		a,d,c		MITSUBI	JAP		59
CCM		a		HUS	JAP	1985	93
DR	C	f				1984	20
MR	T	a,b,d,f		PU	US	1974	1,48
		a,d,f,g	x	PISA	ITALY	1986	66
DTa		a,d,e,f,g	x	KOMATSU	JAP	1985	131
AP	C	f		MIT	US	1982	148
		a,b,c,f		IFW-TU	WG		59
		a,b,c,f		PERA	GB		59
		a,c,f		PSST	WG		59
		a,b,c,f		TOGM	EG		59
MR	EMC	d,e,g	x	UO TWENT	NETH	1986	78,89
		a,c,f		IPAS	WG	1981	22,59
		a,c,d,f		SAG	WG	1981	59
SPC	GCP	a,d		PU	US	1982	29
ES	T	a,c,f		VPT	US	1982	23
		a,c,f		IOT	HU		59,165
DTa	T	a,f,g	x	(WIRTSCH	WG	1981	50
	T	f		IPA-TUS	WG		59
DTr		a,b,c,d		WDEC	US	1980	138
DM	DR	a,c		CAM-I	US	1984	130
MR	T	a,c,f		PSU	US	1982	52
DTr		a,b,c,d,f		BIME	CHINA	1987	87
DTr		a,d,e,f		BIA & A	CHINA	1986	87
DTr		a,b,c,d,f		BIT	CHINA	1986	87
DM	C	a		UMIST	GB	1977	50
		a,d		Elliot	US	1974	108
DR		a,d,f		PSU	US	1987	98
		a,b,c		SAABS	SW		59
SP		a,c,d		LOCKHEED	US	1978	32,70,75
		a,c		CIMB	BUL		59
DTa		f		UMIST	GB	1981	56
SP	T&C	a,f		GE/METCU	US	1977	163
					HU		165

TABLE I

NO.	FULL NAME	Acronym	Approach	Computer Hardware	Rotational	Non-rot.
1.		ABTOTPZK	V		X	
3.	Allie Chalmere Univation DATA	ACUDATA/Univation	V	IBM-370	X	X
4.	Advanced Integrated Mfg. System	AIMS	V			
13.		AUTO-PROGRAMMER	V		X	X
20.		BMW	V	Proprietary	X	X
24.		CAP	V	Mainframe		X
27.	Computer Aided Process Planning	CAPP	V	IBM 370/168	X	X
36.		CODE	V		X	X
39.		COPICS	V		X	X
63.	Interactive Process Planning and Specs.	IPROS	V	Mainframe	X	X
72.	MCdonnel douglas AUTOmation	MCAUTO	V	Proprietary		
74.	Metal Institute Automated Process Planning	MIAPP	V	Mini-Comp	X	X
76.		MICROCAPE	V		X	X
79.		MIPLAN	V	PDP-11	X	X
80.		MIPLAN/MIPREP	V	Mini-Comp	X	X
82.		MITURN	V	Mini-Comp	X	
85.		MULTICAPP	V	Mainframe	X	X
86.	National Engineering Laboratory	NEL	V	Mainframe	X	X
87.	On Line Planning System	OLPS	V	Mini-Comp	X	X
89.		PI-CAPP	V	Mainframe	X	X
102.		SAPT	V			X
110.		SYSTEM AV	V		X	X
112.	Manufacturing data base	TARAMAN	V	Mini-Comp	X	X
113.	Tooling Information DirectorY	TIDY	V	LSI 11/23		X
115.	TOnJI university Comp. Aided Process plan'g	TOJICAP	V	Micro-Comp	X	X
117.		TRAUB	V		X	X
121.		VARGEN	V		X	X
123.		VIOTH	V		X	
11.		AUTODAK			X	
51.		FAUN	-	Mini-Comp		X
52.		FFS	-	Mini-Comp		X
57.		GETURN	-	Mainframe	X	
60.		HICLASS	-	Proprietary		
65.		KOMATSU		Proprietary	X	X
69.	MAnufacturing DEcision MAking	MADEMA	-		X	X
93.	PRAUTO-Turning system development	PRAUTO	-		X	
108.	Software Package Aided the Mfg w/NC m\c tools	SPAM	-		X	
111.		SYSTEM RW	-		X	X
125.	Workshop Programing Systems	WPS	-		X	

205

*S. Gouda, "Micro. Knowledge-Based Computer Aided Process Planning for Rotational Parts (MICAPP)," Doctor of Engineering Thesis, University of Detroit, April 1989.

Column groups — **INPUT**: Part Description (Cad, Code, Other) and Mode (Interactive, Other); **PROCESS**: Logic Type, Selection Criteria; **OUTPUT**: Automation Features, Interface with CAM; **DEVELOPER**: Name, Country, Known year, Reference No.*

Cad	Code	Other	Interactive	Other	Logic Type	Selection Criteria	Automation Features	Interface with CAM	Name	Country	Known year	Reference No.*
						T&C	a,c,f		CIMB	BUL		59
	X		X		SP		a		ALLIS-CH	US	1972	41
									MDAC	US	1977	107
						T&C	f		OB	WG		59
	X				SP				BMW	WG		59
	X				PP		a		LGC	US	1963	22,165
	X				SP		a,c		CAM-I	US	1976	100
	X		X	X			a		MDSI	US	1976	105
	X		X	X			a		MTUF	WG		59
	X		X		SP		a,g	x	NTH-SINT	NOR	1987	127
									MDAC	US	1977	107
	X		X	X	SP		a		OIR	US	1979	22
		X	X		MR	T	f		MARLOW	GB	1981	50
	X		X	X	SP		a		GE/OIR	UA	1979	97,137
	X		X	X	SP	T	f		TNO-M	NL	1980	59
	X		X	X	SP		a,c,d,f,g	x	OIR	US	1971	133,169
	X			X	SP		a,b,c		OIR	US		59
	X		X		SP		a		NEL	GB	1974	50
	X		X		SP		d		BOEING	US	1978	18
	X		X	X	SP		a				1980	22
							a,c		MEB	YUG		59
	X				SP				MICRODAT	WG		59
	X		X	X	SP		a		FORD	US	1981	151
	X		X		SP		d		UOS	GB	1984	33
	X		X		SP		a,b,c,d,f		TONGJI	CHINA	1982	87,178
	X				SP_		a		TRAUB	WG		59
	X				SP		a,c		CRIF-KUL	BUL		59
					SP		a		VOITH	WG		59

Cad	Code	Other	Interactive	Other	Logic Type	Selection Criteria	Automation Features	Interface with CAM	Name	Country	Known year	Reference No.*
		X	X		DTa	T	a,c,f				1976	
		X					f,g	x	TUB	HU		59
		X					f,g	x	CAI	HU		59
		X					d,f,g	x	GE	US	1975	22
									HUGES	US		161
						T	c,d,e		KOMATSU	JAP		59
		X			ES	DR			MIT	US	1985	30
		X					g	x	CAM-I	SW	1977	50
	X						g	x	(BEDINI,	GB	1978	50
		X					f,g	x	WD	WG		59
X							a,d,g	x	(SPUR)	WG	1981	50

5. Surface finish and tolerance knowledge.
6. Selection of operation knowledge.
7. Cutting power knowledge.
8. Machine tool knowledge.

Selection of Optimal Cutting Conditions

Optimal machining conditions can be obtained based upon the selection of one of three criteria which were adopted by the MICAPP system. Listed in the order of increasing cutting speed, these criteria are as follows (2):

1. Minimum Cost per Component (MCC).
2. Maximum Profit Rate (MFR).
3. Maximum Production Rate (MPR).

The MICAPP System Validation

There are three aspects to be validated in the performance of the MICAPP system. The first focuses on validating the reasoning process during the developmental stage. The second concentrates on the intermediate results prior to the final results. The third addresses the final conclusion after the knowledge base becomes more complete. Experts classify the performance of the expert system at the end of these three aspects as ideal, acceptable, suboptimal and unacceptable (13).

Conclusions

An intelligent computer aided process planning system was developed for rotational parts requiring turning operations. It is generative in nature, interactive and runs on microcomputers. This micro-based Computer Aided Process Planning system is named MICAPP. The MICAPP system addressed the creation of a CAD module. It adopts an expert system approach with production rules for knowledge representation and a backward chaining mechanism of reasoning to resolve conflict among the rules. MICAPP selects and sequences machine operations. It acquires knowledge from databases concerning raw materials, cutting tools and machines available in most small or medium shops. THe MICAPP system is able to determine the optimal sequence of operations, and the selection of machines, cutting tools and cutting conditions. The MICAPP system contains effective editing capabilities for the system maintenance.

References

1. D. K. Allen and P. R. Smith, "Computer-Aided Process Planning," Computer Aided Manufacturing Laboratory, Brigham Young University, Provo, Utah, October 15, 1980.

2. E.J.A. Armarego and R. H. Brown, _The Machining of Metals_, Prentice-Hall, Inc., 1969.

3. L. Alting, _Manufacturing Engineering Processes_, Marcel Dekker, Inc., New York, 1982.

4. G. Boothroyd, _Fundamentals of Metal Machining and Machine Tools_, McGraw-Hill Book Company, 1975.

5. T. C. Chang, "Computer-Aided Process Planning Today and in the Future," IIE, Industrial Engineering Conference Proceedings, Fall 1983, pp. 349-355.

6.   B. K. Choi, "CAD/CAM Compatible, Tool-Oriented Process Planning for Machining Centers," Ph.D. Thesis, Purdue University, December 1982.

7.   B. Colding, L. V. Colwell, and D. N. Smith, "Delphi Forecast of Manufacturing Technology-Manufacturing Management," Society of Manufacturing Engineers, Dearborn, Michigan, 1980.

8.   C. T. Culbreth, Jr., "A Microcomputer Based Part Classification and Coding System to Facilitate Computer Aided Manufacturing Applications in the Furniture Industry," Ph.D. Thesis, North Carolina State University, Dept. of Industrial Engineering, Raleigh, 1984.

9.   B. J. Davies, and I. L. Darbyshir, "The Use of Expert Systems in Process-Planning," Annals of CIRP, Vol. 33/1, 1984, pp. 303-306.

10.   D. Dietz, "Tools for Total Quality," Computers in Mechanical Engineering, July/August 1988, pp. 8-13.

11.   T. J. Drozda, and C. Wick (Editors), _Tool and Manufacturing Engineers Handbook_, Fourth Edition, Machining, Society of Manufacturing Engineers, Dearborn, Michigan 1983.

12.   D. Follette, "Machining Fundamentals, a Basic Approach to Metal Cutting," Society of Manufacturing Engineers, Dearborn, Michigan, 1980.

13.   S. Gouda, "Micro. Knowledge-Based Computer Aided Process Planning for Rotational Parts (MICAPP)," Doctor of Engineering Thesis, University of Detroit, April 1989.

14.   S. Gouda, and K. Taraman, "CAPP: Past, Present and Future," SME Technical Paper MS89-368, presented at the 1989 SME International Conference, Detroit, Michigan, May 1989.

15.   S. Gouda, and K. Taraman, "MICAPP System Development," SME Technical Paper MS89-443, presented at the 1989 SME International Conference, Detroit, Michigan, May 1989.

16.   P. Harmon, and D. King, _Expert Systems, Artificial Intelligence in Business_, John Wiley and Sons, Inc., 1985.

17.   J. Hatvany, "What is Economic, and How Am I to Know?" 19th CIRP International Seminar of Manufacturing Systems, Penn State, USA, June 1987, pp. 5-8.

18.   F. Hayes-Roth, D. Watermen, and D. Lenat (Editors), _Building Expert Systems_, V1, Addison-Wesley Publishing Company, Inc., 1983.

19.   B. Logar, and J. Peklenik, "Computer-Aided Selection of Reference Parts for GT-Parts Families," 19th CIRP International Seminar on Manufacturing Systems, Penn State, USA. June 1987. pp. 131-136.

20.   Metcut Research Associates, Inc., _Machining Data Handbook_, 3rd Edition, Vol. 1 and 2, Metcut Research Associates, Inc., Cincinnati, Ohio, 1980.

21.   D. A. Milner, "The Use of Decision Table Logic in Manufacturing Systems," International Journal of Production Research 15/17, 1977.

22. Dana Nau, "Expert Computer Systems, and Their Applicability to Automated Manufacturing," US Department of Commerce, National Bureau of Standards, NBSIR 81-2466, February 1982.

23. C. Townsend, and D. Feucht, <u>Designing and Programming Personal Expert Systems</u>, TAB Books Inc., 1986.

24. H. E. Trucks, <u>Designing for Economical Production</u>, Society of Manufacturing Engineers, Dearborn, Michigan, 1974.

AN EXPERT SYSTEMS APPROACH TO RULE-BASED

DYNAMIC PROCESS PLANT SCHEDULING

W. Marcus Sztrimbely, Peter J. Weymouth and Tony Ponzo

Advanced Dynamic Systems Inc.
8130 Sheppard Ave., East
Suite 200
Scarborough, Ontario
M1B 3W3

Abstract

The secure operation of a complex industrial process, such as a Steelmaking
Meltshop or Nickel Smelter, depends almost entirely on a harmonic working
relationship between the human operator and the real time process control
environment. The introduction of sophisticated real time monitoring systems
and faster than real time decision making modules, requires a new approach
to dynamic process plant scheduling and production optimization under varying
operating constraints. Based on the proven Expert Systems Methodologies of
Knowledge Engineering and Rule-Based Logistics, The ADSI Scheduling Advisor
has been successfully implemented on the shop floor of several major base
metals producing facilities. An overview of several successful projects will
be presented in an attempt to demonstrate how the new and emerging science
of Artificial Intelligence can be successfully integrated with the time
tested 'rules of thumb' operating practice.

Expert System Applications in
Materials Processing and Manufacturing
Edited by M.Y. Demeri
The Minerals, Metals & Materials Society, 1989

Introduction

Industries have been scheduling their manufacturing processes for many years.
The common approach being followed is that of having an expert in plant
operations, usually a senior operator, draw up a schedule of tasks for each
day or shift, making use of a pen, paper and his intuition. The job of the
production scheduler is to plan when each required task should be done, what
equipment will be used and approximately how long the task should take.
Optimizing the schedule in terms of minimizing costs and maximizing
throughput, in even the simplest of plants, is not an easy task.

The ADSI Scheduling Advisor is a computerized scheduling system,
incorporating Advisory Intelligence capabilities based on proven Expert
System Methodologies (rule-based) and extensive data processing. The
scheduling system is designed to reside on the shop floor and to run on a
PC-AT computer or mini. Automatic data transfer between the scheduling
system and the Process Control Environment is by means of an extremely user
friendly window interface. The decision making modules make use of forward
and backward reasoning, conditional reasoning algorithms and sophisticated
search routines in order to optimize the schedule of operations. All
decision making is based on the Artificial Intelligence approach referred to
as 'State of Affairs' in the literature.

The level of operational detail is so refined that it takes into
consideration the impact on production of having an average or better than
average operator controlling various pieces of process equipment. The
effects of equipment maintenance, both scheduled and unscheduled, shift
changes and resource availability are taken into account when producing the
schedule. The impact of environmental constraints imposed due to emission
limitations has been incorporated into the scheduling system through the
'RAINMAKER' module.

The following sections of this paper will briefly address the scheduling
system architecture and capabilities as demonstrated through the successful
implementation of The ADSI Scheduling Advisor into the processing plant
environment.

Scheduling System Architecture

The Scheduling System Architecture can be divided into three distinct
components: i) Interactive User Friendly Window Interface, ii) Advisory
Intelligence Modules, and iii) Graphic Colour Display, Animation and Summary
Modules.

<u>Interactive User Friendly Window Interface</u>. In order to facilitate the
initialization of the scheduling system to the current status of the plant
(prior to linking the system into the Process Control Environment) or for
acknowledging the data transferred to the scheduling system, a series of
interactive windows are presented to the user.

The windows maintain a consistent colour scheme so that information from data
bases (not to be changed by the user through the windows), the process
control environment and required manual input are indicated clearly to the
user. The amount of information presented on each window is minimized so as
not to overwhelm the operator with a cluttered screen of numbers. Depending
on the amount of detail required, a number of sub-windows may appear to
present the operator with a well defined 'work sheet' where such information
may be entered.

212

The windows appear in succession automatically, in a predefined sequence, as subsequent windows will appear with certain information already displayed based on upstream information and rules. This feature has been incorporated in order to assist the operator and to provide guidance. The operator may overrule the recommendations or accept them.

During the creation of the schedule, a 'snap shot' of the shop floor is stored at predefined intervals of time. This feature enables the operator to select specific time slots in the schedule at which to enter modifications to the operating constraints without having to enter the initial condition for each piece of equipment. This feature provides a time savings in using the system prior to linking it to the process control environment or in executing 'What-If' scenarios.

<u>Advisory Intelligence Modules</u>. The decision making logistics and operating rules followed in the daily operation of the plant are entered into the Advisory Intelligence Modules, in the form of 'if - then - else' statements. An example of this rule-based structure is seen in the tapping of a ladle of furnace matte for a converter, as follows.

A converter has requested a ladle of matte be tapped and the Furnace Module has accepted the request. Prior to tapping out the ladle of matte the following rules are executed in the Furnace Module.

- if shift change is occurring, delay matte tap, else

  - if lunch break is occurring, delay matte tap, else

    - check if converter still requires this ladle, and if not cancel matte tap request and search for next task for furnace, else

      - locate available furnace with an empty matte ladle in place, that is not in matte launder maintenance and whose matte depth is above the minimum matte depth.

- if no furnaces are available and meet requirements, delay matte tap, else

  - select furnace with greatest matte reserve and tap out ladle of matte from furnace for this converter and adjust furnace matte reserve.

The modularity of the Advisory Intelligence Modules of The ADSI Scheduling Advisor enables a trained operator to easily enter new operating rules or modify existing rules. It is this feature which allows the system to be 'educated'. The result of this 'learning' is an enhancement in the level of sophistication of the decision making process and a better overall scheduling system.

The Advisory Intelligence Modules have been implemented in the past in FORTRAN-77, as this is the most common computer language in use in the industrial environment of application. Research has been conducted by Advanced Dynamic Systems Inc. into the use of a traditional A.I. language for the Advisory Intelligence Modules. A brief summary of this research will now be presented.

The inherent limitation in using a traditional procedural language such as FORTRAN, is that the knowledge contained in the Advisor must be hard coded in the program. Whereas a traditional Artificial Intelligence language allows for the separation of the control logic of the program from the system knowledge used to model the operating practice.

In the FORTRAN Advisor, the rules are executed as 'when event x is complete, schedule event y to commence'. The A.I. approach modifies this logic pattern in that the 'inference engine' will search the data bases for rules that can be executed under the present 'State of Affairs' within the shop. The Advisory Module will then select based on heuristics the rule(s) which will be 'fired'. The end result produced by each Advisor will be the same, provided a sufficient number of 'if - then - else' statements are contained in the FORTRAN version and the order in which the statements are evaluated is dynamic, i.e. the path through the statements is dependent on the 'State of Affairs'.

Additional advantages arising from the use of traditional A.I. languages are:

1) the ability to easily backtrack to any decision point and re-evaluate the decision made, thus allowing the Advisor to investigate multiple branches in the 'tree' of possible schedules,

2) the modification of the knowledge base of rules is independent of the control logic and is therefore much more straight forward,

3) the use of natural language interfaces for modification of rules coupled with consistency checking of the data base capabilities, increases the user friendliness of the system to a computer illiterate user, and

4) the ability of the system to add to its own data base or modify existing rules through current A.I. technology, increases the system's ability to 'learn' as well as to be 'educated' based on past experience.

One point should be made at this time. Traditional A.I. techniques are without a doubt the way of the future. Their successful implementation, will only be possible if the introduction of this technology is gradual and the applications are selected with caution.

<u>Environmental Constraints Module 'RAINMAKER'</u>. In the daily operation of a Copper/Nickel Smelter the SO_2 emissions are of grave concern. Present and future laws are placing tighter and tighter restrictions on the level of emission considered as acceptable. Two alternatives to meeting these constraints are available. The smelter may be upgraded with modern equipment and gas handling systems or the production may be reduced to a point where emission rate is met.

The modernization of a smelter does not occur over night, and as such a means of operating during the upgrading must be found. The Environmental Constraints Module 'RAINMAKER' has been implemented in the scheduling systems of several ADSI clients. RAINMAKER will maximize the smelter throughput under the imposed environmental constraints.

The RAINMAKER module is activated on an hourly basis, at which time the 'State of Affairs' within the shop is evaluated and various pieces of processing equipment are taken 'out-of-stack', in the case of converters, or shut down in the case of roasters or furnaces. The one hour time step is found to be satisfactory as a typical converter blow cycle is approximately of this duration.

<u>Graphic Colour Display, Animation and Summary Module</u>. The results from the
scheduling system are available in various formats:

- colour graphic schedule (terminal display and printed),

- colour animation terminal display, and

- summary reports and charts.

The terminal display of the graphic schedule is in colour as this allows for
easy cross-referencing of material flow from source to sink, e.g. Furnace
Matte to a Nickel Converter or the tracking of specific grades or heats of
steel within a Meltshop. The system provides the user with the option of
viewing the graphic schedule in various window widths of time. The user can
scroll forward or backward through the display as well as select the
equipment schedule to be viewed, e.g. furnace schedule, crane schedule,
caster schedule, etc.

The printed copy of the schedule is available either in colour or black and
white. The schedule format has typically been designed in the past to fit on
a piece of 8½ x 11 inch computer paper. This size restriction allows the
schedule to be neatly folded to comfortably fit into a shift foreman's
pocket.

The output from The ADSI Scheduling Advisor may be sent to The ADSI Animation
Module. This stand-alone module allows the plant layout to be drawn on a
colour graphics terminal. The schedule of processing operations is then
displayed as a dynamic animation, e.g. flames out of the mouth of a blowing
converter, hot metal flow as billets from a caster, crane movements, etc.

Two additional features have resulted from implementing The ADSI Animation
Module. The animation display has been adapted to function as an interface
to a plant-wide management information system. Rather than sorting through
screen after screen of data, the status of all process equipment is available
through an information window system.

The second feature is the ability to move the physical location of equipment
on the screen through a 'CAD' style mouse interface. The new location or any
additional equipment is automatically fed to the scheduling system and the
impact of the change may be determined. The animation display capability
enables the effect of the changes in operating practice, equipment
availability, crewing, etc., to be demonstrated dynamically.

A further benefit of this animation module is in operator training. The
impact of various decisions made by an operator on the overall operation are
displayed for review and evaluation, literally within minutes.

Summary reports are available for printout in a format identical to the
existing daily reports and provide the user with key operating information.
The information is also available in a statistical format of run charts,
power curves, histograms, etc. Included in the analytical package is the
ability to superimpose various pieces of equipment on the same chart or curve
for ease of comparison. All items are presented in a consistent colour
scheme to aid in the comparison. Various presentation formats are available,
each designed for a specific application, to highlight the desired
information.

<u>Proven System Capabilities</u>. The ADSI Scheduling System has been used for
five different applications:

    i) overall production planning and forecasting,

   ii) detailed production troubleshooting,

  iii) operator training,

   iv) engineering simulation tool for analysis of different 'Capital Cost'
       alternatives, and

    v) Research & Development tool for answering the numerous 'What-If'
       questions that previously had to be based on best 'guesstimates' for
       the various scenarios being investigated.

The scheduling system is used by the Plant General Foreman, as a predictive
tool for planning production over a twenty-four hour period, based on the
equipment and manpower he will have available. The scheduling system will
advise the General Foreman as to the best time for performing daily
maintenance activities, so as to reduce the negative impact these activities
have on production throughput. The system is used by production management
for long-term planning and budgeting, (long-term simulation).

The scheduling system is used by shift foremen as a troubleshooting tool, in
order to evaluate various scenarios to smooth out a production upset as
quickly and as economically as possible.

The system has been used extensively by plant operations as a visual training
tool for both seasoned and rookie operators. The ability of the user to
quickly produce schedules and to view the results in both graphic and
animated formats, allows the operator to see the impact on plant throughput
and profitability of his decisions.

The capability of using the scheduling system as a simulation tool has been
the key reason for the exhaustive use by both the Engineering and Research &
Development groups. The system has been used to investigate and quantify the
overall impact on production, resulting from both minor and major process and
equipment changes.

The results generated and presented in the form of a colour, cross-referenced
material flow schedule has increased the understanding of both production and
support groups, of the complex and intricate relationships between various
decisions, operations and equipment. This new understanding and the ability
of the program to answer many 'What-If' questions in a qualitative manner,
has resulted in increased plant throughput and profitability with minimal to
no capital cost expenditures.

Conclusions

The ADSI Scheduling Advisor has been developed for use by non-computer
trained personnel. Because of the Expert Systems approach followed in the
development of these systems, the existing knowledge of plant operations is
captured within the rule-based structure of the system and the personality of
the plant is reflected in the production schedules that are produced.

The modular structure of the system enables the system to 'learn' over time,
to reflect changes in operating practice or conditions. The major
contribution to the acceptance of The ADSI Scheduling Advisor comes from the

216

user friendliness that the system has incorporated. Past installations have
indicated that any person familiar with the plant operation, can generate
reliable and realistic schedules after less than one hour of training.

Acknowledgements

The authors would like to express their sincere appreciation to the companies
who have implemented The ADSI Scheduling Advisor within their processing
plant environment, and to acknowledge the support of the National Research
Council of Canada in the development of the basic scheduling system
libraries.

The authors would also like to thank the TMS-EPD SAMP Committee and TMS for
the opportunity of presenting this paper.

Productivity Improvement Through

Machine-Diagnostic Expert Systems

H. Nivi,   I. A. Nagisetty

Ford Motor Company
Manufacturing Development Center
24500 Glendale Ave.
Detroit, MI 48239
U.S.A.

T. M. Ngo

Ford Motor Company
Essex Engine Plant
1 Quality Way
Windsor, Ont N9A 6X3
Canada

L. D. Bradford

Tocco Inc.
30100 Stephenson Hwy
Madison Heights, MI 48071
U.S.A.

Abstract

Determining the root cause of machine failure plays a critical role in machine uptime and directly impacts productivity. Rapid, accurate diagnosis is key to determining the root cause of failure. Ford Motor Company has developed an expert system to diagnose induction-heating equipment made by Tocco Inc. It combines the expertise of experienced Ford personnel with the knowledge of Tocco's best diagnosticians. Plant personnel at Ford use this user-friendly system on a portable computer. It asks a series of questions and provides possible answers, with comprehensive graphics to assist the user in performing the pertinent tests. The *Tocco Diagnostic Assistant* has provided valuable insight into the development of machine-diagnostic expert systems.

Expert System Applications in
Materials Processing and Manufacturing
Edited by M.Y. Demeri
The Minerals, Metals & Materials Society, 1989

Acknowledgements. The authors would like to thank Messrs C. E. Feltner, J. H. Barry and M. Leary of Ford Motor Co. and M. Hammond of Tocco Inc. for their support and encouragement. Also, they would like to thank Messrs R. Reynolds, S. L. Shmuter, and Y. A. Hamidieh for their technical contribution at the start of this project. And thanks to W. H. Lemay for his contribution in documentation and software upgrade.

Introduction

Ford Motor Company has developed an expert system to diagnose failures in induction-heating equipment manufactured by Tocco Inc. The *Tocco Diagnostic Assistant* is a complete software package which incorporates the expertise of both Ford plant personnel and Tocco diagnosticians. This package finds the root cause of equipment failure quickly and effectively, improving quality and productivity.

The Problem

Ford Motor Company uses induction-heating equipment to harden metal parts. Tocco Inc. manufactures such machines. The machines are very reliable and are used in critical operations at many plants. If an induction-heating machine does fail, the downtime has a major impact on the productivity of an entire plant. Experience shows, moreover, that either the cause of the failure is found within the first 2 or 3 hours or it is not found for several days.

To reduce downtime, the Essex Engine Plant expressed a need for a fast and accurate means of diagnosing failures of its Tocco heating machines. They had some diagnostic expertise in house, so they were not entirely dependent on repair personnel from the manufacturer. However, the in-house experts were not always available, and the failures occur rarely enough that the expertise was not constantly refreshed by use.

There are dozens of these Tocco machines throughout Ford Motor Company. Therefore, a solution to the problem at one plant would provide a substantial benefit to other plants within the Company.

The Machine

The Tocco induction-heating machine creates a magnetic field around the targeted metallic part. Induced electrical currents cause the metal to heat up. The machine has several major constituents: rectifier, inverter, control and logic circuits, hydraulic cooling, as well as electrical and mechanical elements at the workstation (Figure 1). These constituents are only partially independent: an apparent failure in one may, in fact, be a failure in another.

The Construction of the Expert System

Tools

The complexities of the induction-heating machine make it attractive to express the cause-and-effect dependencies of one component upon another as a list of rules. The rules connect each cause with its effects. Backward chaining through the rules leads the diagnosis from the effects (that is, the symptoms of the failure) to the cause.

Therefore, a rule-based expert-system shell was needed. The *Personal Consultant Plus* from Texas Instruments provided several useful features.

1. It is based on the IBM-compatible personal computer, which is a convenient hardware platform for use in Company plants.

2. It permits externally-prepared full-screen images presenting explanatory drawings to the user.

3. It comes from an established vendor, so the prospects for present and future product support are excellent.

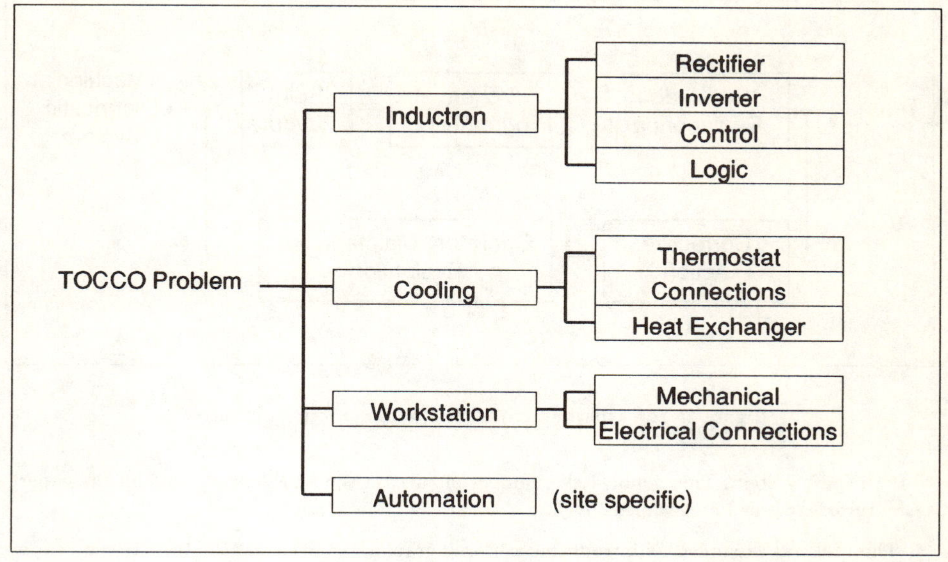

Figure 1: The expert system's structure parallels the structure of the induction-heating machine.

The strategy if applying the expert system was to keep the diagnostic package external to the process equipment. The human technician would be the connection between the induction-heating machine and the diagnostic system. Therefore, there was no need to connect the diagnostic system to the Tocco machine.

Development

After choosing the expert-system development tool, the knowledge engineer familiarized himself with the induction-heating machine. He needed at least a novice's understanding of the machine as a foundation for all the subsequent work and for communication with the machine (domain) experts. The initial sources were operating and maintenance manuals.

Next, the knowledge engineer drew upon the plant experts, the people who actually adjust and maintain the induction-heating machinery. With them, he clearly defined the problem and assimilated the diagnostic and repair procedures followed in the plant. From this, he built an initial prototype system. The plant experts reviewed the prototype for accuracy, although it was still incomplete. (See Figure 2.) End users and the management reviewed the prototype, looking especially at the clarity and convenience of the user interface. Further refinements were made.

The prototype proved that such a machine-diagnostic expert system was both feasible and useful. It was based only on the manufacturer's published operating and repair procedures and on the skill and experience of the responsible people in one plant. To complete the diagnostic system would clearly require deeper, even proprietary knowledge of the equipment.

Therefore, the knowledge engineer approached the manufacturer. Tocco Inc. saw an opportunity to use the expert system internally, including the training of their own repair people. Ford and Tocco reached a shared-technology agreement to complete the development of the expert system jointly and to protect each Company's interests.

The knowledge engineer then added Tocco's in-house expertise and proprietary knowledge

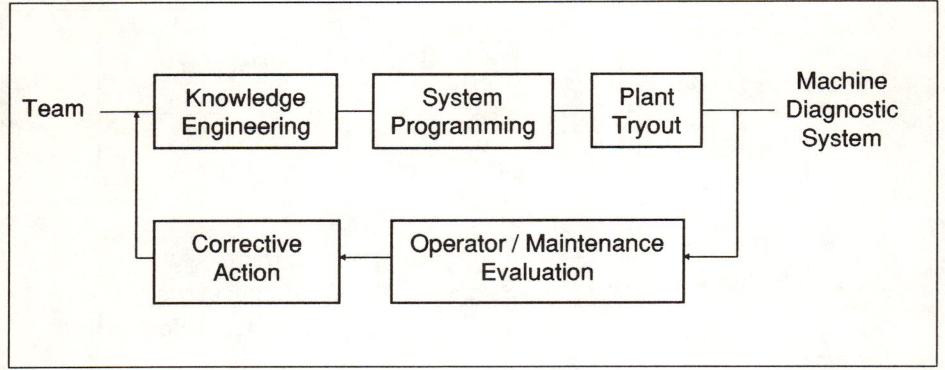

Figure 2: The expert-system development process is iterative.

to the expert system. This doubled the number of rules in the knowledge base, and the expert system was expanded and refined.

The tool was expanded by introducing pictorial graphics to illustrate the tests to perform or to clarify the locations of components (Figure 3).

The difficulty (and, therefore, the cost) of tests influenced the sequence of steps designed into the diagnostic procedure. For example, testing certain components in the inverter requires partially dismantling that unit. Testing comparable components in the rectifier is much easier, and correcting a rectifier error may make dismantling the inverter unnecessary. Considerations such as this influenced the sequence of diagnostic steps.

The system was further improved by turning it over to naïve users, technically trained persons without prior knowledge of the Tocco induction-heating machinery. These users do not know the "right" answers to questions, so they follow paths through the rules to dead ends and other unanticipated outcomes. They also spot unclear language and peculiar instructions in the user dialogue. In all, about a dozen experts and non-experts examined and reviewed this expert system.

The result was the *Tocco Diagnostic Assistant*, a machine-diagnostic system which is accurate and effective as well as clear and robust.

Deployment and Commercialization

The *Tocco Diagnostic Assistant* was placed in one plant and quickly proved its value. A second contract was negotiated between Ford and Tocco, for Tocco both to handle the distribution to Ford plants and to sell the *Tocco Diagnostic Assistant* to other companies owning the same induction-heating equipment. There are about a thousand such installations worldwide.

Results

Although the *Tocco Diagnostic Assistant* has been available to plants only since late 1988, the users have been very enthusiastic. The mean time to repair (MTTR) has been reduced substantially. The thoroughness of the program assures that the true causes of the failures are being found, thus lengthening the mean time between failures (MTBF).

The results were so satisfactory that a second version of the program was written. The first version had been written for the "modular" model of induction-heating machine. The "compact"

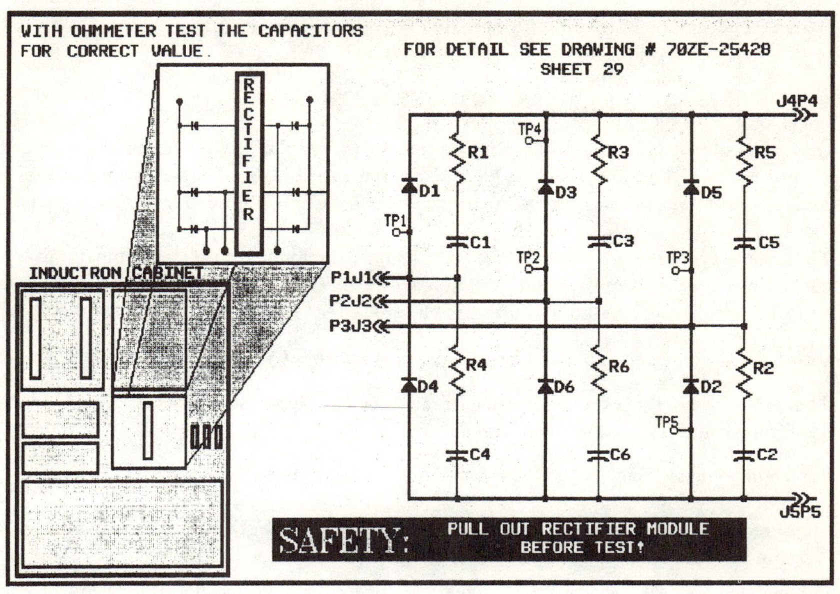

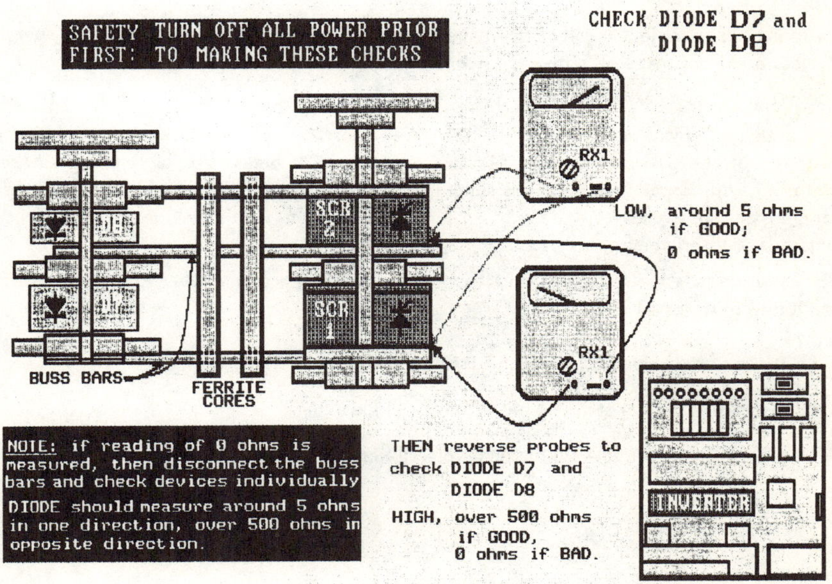

Figure 3: At each point in the diagnosis, full-screen illustrations are available on request. Some provide aid in locating components. Others explain how to perform specific tests.

model is similar in principle but different in organization and physical structure. A *Tocco Diagnostic Assistant* for it was developed quite quickly by using about 40% of the modular model's rules. This second version is also rapidly proving its value.

Conclusions

Two models of Tocco induction-heating equipment ("modular" and "compact") have critical roles in Ford plants. Downtime must be kept as brief as possible. Two expert systems, the *Tocco Diagnostic Assistants*, have been written to speed repair after a failure. They do so because:

1. The expert system's rules put the repair manuals, the manufacturer's diagnostic practices, and plant experience into an automated, easy-to-access form. Thumbing through manuals and phone calls to the manufacturer are eliminated.

2. Responsible plant personnel can initiate diagnosis without waiting for The Expert to arrive.

3. The accuracy of the expert systems streamlines the diagnostic process. There are fewer blind alleys and less "experimental repair."

4. Foresight makes efficient use of the repair technician's time.

Besides reducing the mean time to repair, the programs lengthen the mean time between failures by finding the root cause of the problem the first time.

The construction of these expert systems was an experiment in the application of expert-system techniques to real-world problems. The Tocco induction-heating machines were chosen for this experiment because the machines were of substantial complexity and of key importance in Ford plants.

Involving the manufacturer had two benefits: (1) bringing the maximum level of expertise to the project and (2) providing a means to commercialize the product.

The overall plan worked very well, and it serves as a model for similar projects in the future.

One addition, however, is desirable. There are no precise records of repair times before the introduction of the *Tocco Diagnostic Assistant*. Measure of benefits is, accordingly, based on anecdotal information. A future project should collect maintenance data from the time before the new diagnostic tool is used—at the very least while the system is in development. Then accurate comparisons of improvement may be made.

The final diagnostic systems exhibit features missing from many other expert-system applications found in industry today:

1. They are marketable and are commercially available to end users.

2. They capture the expertise of more than a single expert.

3. They serve as a valuable supplement to technician training.

Subject Index

Author Index